# THE EXOTIC
# AMPHIBIANS
# AND
# REPTILES
# OF FLORIDA

# THE
# EXOTIC
# AMPHIBIANS
## AND
# REPTILES
## OF FLORIDA

**WALTER E. MESHAKA, JR.**
Zoology and Botany Section
State Museum of Pennsylvania
Harrisburg, Pennsylvania

**BRIAN P. BUTTERFIELD**
Department of Biology
Freed-Hardeman University
Henderson, Tennessee

**J. BRIAN HAUGE**
Department of Biology
Pennsylvania State University Hazleton
Hazleton, Pennsylvania

KRIEGER PUBLISHING COMPANY
Malabar, Florida
2004

Photographs used with permission of Richard D. Bartlett; Suzanne L. Collins, The Center for North American Herpetology; Joseph T. Collins; Thomas E. Lodge; Samuel D. Marshall; Bradley M. Stith; and R. Wayne Van Devender.

Original Edition 2004

Printed and Published by
**KRIEGER PUBLISHING COMPANY**
**KRIEGER DRIVE**
**MALABAR, FLORIDA 32950**

Library of Congress Cataloging-in-Publication Data

Meshaka, Walter E., 1963-
   The exotic amphibians and reptiles of Florida / Walter E. Meshaka, Jr.,
Brian P. Butterfield, J. Brian Hauge.
      p. cm.
   Includes bibliographical references (p. ).
   ISBN 1-57524-042-4 (alk. paper)
      1. Amphibians—Florida.   2. Reptiles—Florida.   3. Exotic animals—Florida.
   I. Butterfield, Brian P.   II. Hauge, J. Brian.   III. Title

QL653.F6M37 2004
597.8′09759—dc21

   10   9   8   7   6   5   4   3   2                                     2003054575

# DEDICATION

WEM- To the memory of Robert L. Cox, fellow herper and friend.

BPB- In memory of my grandfather, P.C. Dorris, a man responsible for nurturing my early love of nature.

JBH- To my wife, Joy, for everything. To my parents, Jim and Betty Hauge, and Joy's parents, David and Marrelyce Seaman, for their support. To John Haertel, Craig Guyer, Bob Mount, and Jim Dobie for teaching me herpetology in the classroom and in the field.

# CONTENTS

# ACKNOWLEDGMENTS

For assistance, be it in the field, the laboratory, or in conversation, we wish to acknowledge the following people: R.D. Bartlett, O.L. Bass, Jr., P.D. Bedsole, E. Blankenship, E. Blind, J. Burgess, K.L. Butterfield, J.T. Collins, J. Dippel, R. Ehrig, B.W. Emanuel, B. Ferster, J. Fischer, G. Folkerts, J. Freezer, C. Guyer, G. Harrison, D. Holley, D. Joyce, R. Kilhefner, J.D. Lazell, Jr., J. Lewis, S.D. Marshall, E. Mayr, E. McDuffee, S.B. McGee, K. Nicholson, L.D. Ober, J. Provenzale, W. Reynolds, W.B. Robertson, Jr., G.W. Sabbag, R. St. Pierre, S. St. Pierre, K. Semuta, K. Skocik, A. Solinski, B. Stith, K.A. Wallick, T. Walsh, L.D. Wilson, and G.E. Woolfenden.

# INTRODUCTION

The presence of exotic amphibians and reptiles in Florida has drawn attention from herpetologists since the late 1800s. Herpetologists have documented increasing numbers of exotic amphibians and reptiles in the state beginning with the introduction of the greenhouse frog (*Eleutherodactylus planirostris*) in Florida over 125 years ago (Cope, 1875). Key treatments include those by Carr (1940), Duellman and Schwartz (1958), King and Krakauer (1966), Wilson and Porras (1983), and Butterfield and colleagues (1997). However, most treatments of this segment of Florida's herpetofauna generally focused more on the geographic distributions and histories of introductions than on the natural histories of the species (Carr, 1940; Duellman and Schwartz, 1958, Wilson and Porras, 1983; Ashton and Ashton, 1988a, 1988b, 1991; Bartlett and Bartlett, 1999).

Since the introduction of *E. planirostris,* Florida has continued to be an area where exotic amphibians and reptiles have become established. In fact, Florida contains more exotic amphibians and reptiles than any other U.S. state (Wilson and Porras, 1983; Butterfield et al., 1997). This influx of exotic species has resulted in some notable changes. For example, Florida contains more exotic than native lizard species. Additionally, some of the exotic species have become predators and/or prey of native species. Exotic species have also been implicated as competitors of native species; however, this allegation has generally not been studied.

Studies addressing the consequences of the establishment of exotic amphibians and reptiles in Florida should be conducted; however, the basic biology of these species needs to be understood before these types of studies will be of full value. To date, most authors who have reported the distributions and histories of introductions of the established exotic species have rarely discussed the biology of the species. Therefore, a book has been needed not only to update the list of exotic amphibians and reptiles in Florida, but also to provide a progress report on new and published natural history information for each of what is now 40 documented species established in Florida. The state's 67 counties are shown in Figure 1.

Although Dade County recently underwent a name change to Miami-Dade County, in the interests of familiarity to a broad audience that extends beyond Florida we are retaining the better-known name of Dade County in the text.

This book updates the list of exotic amphibians and reptiles in Florida and provides available natural history information for each species in the form of species accounts. Specifically, we present the history of introduction, geographic distribution, and natural history for each established species. We also provide a list of species whose population status in Florida is uncertain. The afterword, by

1

**Figure 1.** Florida's 67 counties.

Walter E. Meshaka, Jr., "The Human Role in the Colonization of Species in Florida," discusses the role of exotic species in the context of conservation.

During our research on exotic species and while assembling this book, we identified three trends associated with the taxonomic, temporal, and spatial patterns of colonization by these 40 species.

1. Most of these species are lizards (32 species), followed distantly by frogs and toads (4 species), snakes (2 species), turtles (1 species), and crocodilians (1 species). Among the lizards, most are either geckos (11 species) or anoles (8 species).
2. The rate of exotic species accumulation has increased over time, especially in the past 30–40 years (Figure 2).
3. Areas most frequently colonized are in urban southern Florida (Figure 3).

We have listed the three trends in the beginning of this book to encourage readers to view the species not just as individuals, but instead as an entire new fauna within Florida.

You will likely note in the *History of introduction and current distribution* section of the species accounts that many of these species became established when they were accidentally or intentionally released. We must point out that release

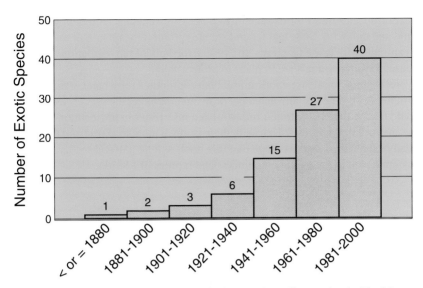

**Figure 2.** Accumulation of exotic amphibians and reptile species in Florida.

**Figure 3.** Distribution of exotic amphibians and reptiles in Florida.

of exotic species into Florida (and most other U.S. states) is both biologically un-sound and unlawful. State and federal statutes prohibit the release of exotic plants and animals with few exceptions. In Florida, the bulk of the regulations regarding release of exotic animals are in place to protect aquatic ecosystems. Nevertheless, anyone wishing to release exotic amphibians or reptiles for whatever purpose (for example, to be rid of an unwanted pet or to attempt to get the species established) should realize that this action is unlawful and may result in prosecution leading to fines and/or imprisonment. That said, there is much more to learn from the species already established, and we hope that this book will be a useful resource for per-sons interested in the exotic amphibians and reptiles of Florida.

# SPECIES ACCOUNTS

Our objectives for each account are to summarize the literature concerning geographic distribution, to discuss the origin of introduction, agent(s) and rate of dispersal, and to provide natural history information from Florida. It is our intention to present these data only for exotic species that are established in Florida. For this reason, we have adopted the following criteria regarding establishment:

- The species is recorded in the form of a voucher specimen or photograph. This criterion ensures that species identifications can be confirmed.
- The species shows evidence of reproduction (other than in captivity). In most cases, gravid females, hatchlings, or several size-classes have been collected. In a few cases, however, we have included species for which there are no reports of gravid females or neonates. The Puerto Rican coqui (*Eleutherodactylus coqui*) is an example. We have included this species in the accounts because of the persistence of chorusing groups from the same area for more than one decade, leading us to believe that this species is reproducing in Florida.
- The colony has been in existence for at least one-generation time (the amount of time it takes for the first offspring of a colonizing species to reach maturity and reproduce). For long-lived, slowly maturing species such as the red-eared slider (*Trachemys scripta elegans*), this could be as long as 8–10 years. For short-lived, rapidly maturing species such as the house gecko (*Hemidactylus frenatus*), this could be less than 1 year. While taking generation time for each species into account is preferable to picking some arbitrary time period (such as 5 years or 10 years), generation time is not known for all introduced species in Florida. Therefore, we considered populations with several size- or age-classes to have met this criterion.

The taxonomic arrangement of the species accounts follows the organization of Pough et al. (2001) to best reflect our current understanding of relationships among these groups. Species are listed in alphabetical order within higher taxonomic groups. Higher taxonomic groups are listed in alphabetical order in Appendix A for convenience. Common names of amphibians and reptiles follow Collins and Taggart (2002). We use common names provided by Frank and Ramus (1995) for species not listed in Collins and Taggart (2002). Other vernacular names are also provided. A cross-reference of scientific and common names of animals mentioned in this book is provided in Appendix B.

Three symbols appear on the distribution maps that accompany each species account. Stars represent literature records through December 2001 of a species from specific geographic localities. Solid circles represent unpublished records that are supported by voucher specimens or photographs deposited in the following:

- Archbold Biological Station Vertebrate Collection (ABS), Lake Placid, Florida
- Auburn University Museum (AUM), Auburn, Alabama
- Everglades Regional Collections Center (EVER), Homestead, Florida
- Florida Museum of Natural History (FMNH), Gainesville, Florida
- University of Kansas (KU), Lawrence, Kansas
- University of Michigan Museum of Zoology (UMMZ), Ann Arbor, Michigan
- National Museum of Natural History (USNM), Washington, D.C.

Open circles represent reports of reliable observations without a voucher specimen or photograph. Gray shading indicates region-wide distribution of a species, and only published county records are included in shaded areas. Because of interest in insular dispersal, records of species on individual islands of the Florida Keys are shown.

Voucher specimens will be referenced in the text as per the acronyms listed above.

For measurement purposes, the following measurements and abbreviations are used. For lizards, snakes, and crocodilians, total length (TL) is the distance from the tip of the snout to the tip of the tail. For frogs and toads, snout-vent length (SVL) is the distance from the tip of the snout to the posterior margin of the vent. For lizards and snakes, SVL is the distance from the tip of the snout to the posterior edge of the scale or flap of skin that covers the vent. For crocodilians, SVL is the distance from the tip of the snout to the anterior border of the vent. When measuring turtles, carapace length (CL) is the midline length of the carapace or upper shell of a turtle. For our original data, a mean value is followed by ± standard deviation, range of values, and sample size (N). The natural history portion of the species accounts (*History of introduction and current distribution, Habitat and habits, Reproduction, Diet,* and *Predators*) presents the most current information from the literature and field observations of these species in Florida. Collectively, the species accounts provide a snapshot of the current status of each species. A section on *Species of Uncertain Status* documents those species whose establishment and ability to persist in Florida are questionable or whose colonizations have failed.

With respect to collection and observation of these species, we cannot overstate the importance of recognizing and honoring property rights and concerns of landowners. Many exotic species are restricted to buildings and/or surrounding vegetation, and landowners may even encourage their presence. While working in the field, we always attempt to work with property owners, asking them not only for their permission to collect, but also what they know about the animals we are collecting. Many times we gain greater knowledge about a species by stopping to ask questions than we could possibly have gained by simply collecting. Clearly, this book would have contained far less information and fewer species accounts had we not had the cooperation of and honored the wishes of landowners who allowed us to observe, report, and collect species inhabiting their properties.

# ANURA: FROGS AND TOADS

## FAMILY BUFONIDAE

*Bufo marinus* (Linnaeus, 1758)

**Common name:** Cane Toad

**Other common names:** Giant Toad, Marine Toad

**Description:** The dorsum is brown with variable pattern. The venter is white with brown specks. In breeding males the dorsum is cinnamon and its skin texture is spinose. This toad has large parotoid glands that extend over the shoulders. Tadpoles are small in body size and uniformly dark brown to black in body color with gray or tan caudal fins. The body appears globular from above and somewhat flattened from the side. The eyes are dorsally placed, the vent tube is dextral, and the spiracle is sinistral. The dorsal and ventral fins of the tail are of approximately equal depth. These tadpoles have two upper and three lower rows of teeth.

**Body size:** The largest male (150-mm SVL) and female (175-mm SVL) were from Lake Placid, Highlands County.

*Bufo marinus* Cane Toad. Photo by W.E. Meshaka, Jr.

**Similar species:** *Bufo marinus* differs from the native oak toad (*B. quercicus*) and southern toad (*B. terrestris*) by having a much larger adult body size and by having large parotoid glands that extend over the shoulders. *Bufo terrestris* has prominent knobs on the posterior portion of the interorbital crests whereas *B. marinus* lacks such knobs.

**History of introduction and current distribution:** *Bufo marinus* is native from extreme southern Texas, south through Central America and into much of tropical South America. It was first introduced into Florida for pest control in sugarcane fields. These introductions consisted of the release of approximately 200 individuals into two Palm Beach County sites (Belle Glade and Canal Point) prior to June 1936 (Riemer, 1958) and release of an unknown number of individuals at Clewiston, Hendry County, prior to 1944 (Oliver, 1949). Subsequent introductions were made at Pennsuco, Dade County, and at Bass Biological Laboratory at Englewood, Sarasota County (Riemer, 1958). Oliver (1949, 1950a) found no evidence of colonization success for any of these introductions and attributed failures to cold weather.

The origin of the first successful introduction of *B. marinus* into Florida appears to have been an accidental release of approximately 100 toads by an animal importer near the Miami International Airport, Dade County, prior to 1955 (King and Krakauer, 1966). Subsequent releases at Pembroke Park, Broward County, and Kendall, Dade County, were deliberate (King and Krakauer, 1966). Duellman and Schwartz (1958) believed that the Dade County population was more likely derived from accidental or deliberate pet releases than from the earlier release at Pennsuco. Soon thereafter, *B. marinus* was reported from the counties of Dade (Neill, 1957; Allen and Neill, 1958; Duellman and Schwartz, 1958), Broward (King and Krakauer, 1966), Palm Beach (Krakauer, 1968), and recently in Pasco (Stevenson and Crowe, 1992a). Populations have been known from Tarpon Springs, Pinellas County; and Tampa, Hillsborough County, since 1956 (Rossi, 1981). On the Florida Keys, Monroe County, *B. marinus* was recorded on Stock Island (Krakauer, 1970) and Key West (Wilson and Porras, 1983) and is still present on those keys.

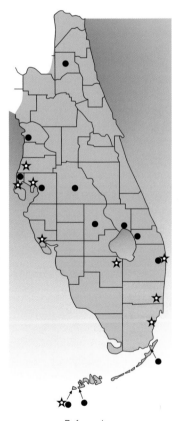

*Bufo marinus*

We have recorded *B. marinus* from Citrus, Highlands, Okeechobee, and Pinellas counties and from Key Largo. The Highlands County (Lake Placid) population has been established since the late 1970s following the development of a residential subdivision within a sandy upland habitat. Presently, the Highlands County population occurs only within this subdivision. The Okeechobee population dates back to the mid-1970s. Both colonies are of unknown origin. The FLMNH lists specimens from Clay, Martin, and Polk counties among its holdings.

**Habitat and habits:** *Bufo marinus* generally occupies habitats that are moist, open, and human-modified. It prefers areas with permanent water nearby but avoids wet prairies and sandy dry areas. It is often found in rural subdivisions, golf courses, and school campuses. All three modified habitats typically possess characteristics attractive to these toads: lights that attract insects for prey, wet areas that provide sites for breeding and hydration, and asphalt and buildings that radiate heat. Although usually found in open areas, this species has been observed in dense stands of Brazilian pepper (Rossi, 1981). No records exist of *B. marinus* being established in natural systems anywhere in Florida, and we have only one report in 1990 of an adult from Anhinga Trail in Royal Palm Hammock, Everglades National Park (Meshaka et al., 2000). Biweekly visits at night (1990–1992, 1995–1996) to Anhinga Trail and other sites in the park and extensive nocturnal visits to those sites from January 1997 to May 2000 revealed no other individuals. This species is, however, very abundant in the surrounding agricultural areas of Florida City and Homestead. That agricultural belt provides *B. marinus* with the relatively open habitat that it seems to prefer and a hydrologically homogenized system that is neither xeric nor permanently inundated with water.

Adults and subadults are primarily nocturnal, whereas metamorphs are diurnal. On warm days following rains, adults and subadults emerge from diurnal retreats to forage. They will also emerge during the day to feed on pungent-smelling food scraps (Rossi, 1983). Although this species is terrestrial, Rossi (1981) once found a *B. marinus* presumably hunting at night 2 m above the ground on a slanted tree. Tadpoles forage in large groups in shallow water. Metamorphs have been observed moving about during the day not far from standing water.

Activity varies among regions. In Miami and Lake Placid, individuals were most frequently seen at night from March through October, although activity occurs continuously throughout the year. Activity in Tampa occurs during March through November (Rossi, 1981). Winter activity is suppressed except near water (Krakauer, 1968), and adults of *B. marinus* from Miami die when exposed to temperatures of approximately 5°C for 96 hours (Krakauer, 1968).

From the late 1960s through the late 1970s, population sizes of *B. marinus* in Dade County were high. However, in the early 1980s, population densities sharply decreased for unknown reasons. Moreover, anecdotal evidence suggests that the average body size is smaller now than in the early 1980s.

Ticks and hookworms have both been found to parasitize *B. marinus*. Some Florida populations harbor the tick *Amblyomma rotundatum* (Oliver et al.,

1993). We found this tick on *B. marinus* from the sites studied by Oliver and coworkers. This parasite was introduced to Florida with *B. marinus* and has been present in southern Florida populations since 1979. This tick is most commonly found on *B. marinus* during the dry season. However, no ticks were detected in a sample of more than 200 individuals collected throughout the year from Lake Placid. Rossi (1981) found *A. americana* and *Dermacentor* sp. on Miami samples of *B. marinus* but none on individuals from Tampa. Ova of the common hookworm (*Aclyostoma caninum*) have been recorded from the feces of *B. marinus* from Tampa, which may be a result of *B. marinus* eating dog feces (Rossi, 1981).

**Reproduction:** Males are sexually mature between 75-mm SVL (Stock Island, Monroe County) and 95-mm SVL (Lake Placid, Highlands County). Females mature sexually between 89-mm SVL (Florida City, Dade County) and 95-mm SVL (Stock Island, Monroe County). The smallest gravid female from our collections is 89-mm SVL. Among adults, males are on average smaller in body size than females (Table 1).

The call of the male is a low drumming. In southern Florida, males are capable of calling throughout the year. We observed this species calling from Jan-

**Table 1.** Mean and standard deviation of body size in mm SVL at sexual maturity of *Bufo marinus* from five locations in Florida.

| Location | Male | Female |
|---|---|---|
| Stock Island<br><br>Monroe County | 99.4 ± 15.7;<br>range = 75–120<br>N = 9 | 121.7 ± 22.5;<br>range = 95–150<br>N = 3 |
| Florida City (downtown)<br><br>Dade County | 99.3 ± 11.9;<br>range = 90–135<br>N = 14 | 107.9 ± 17.6;<br>range = 90–140<br>N = 11 |
| Florida City (agricultural)<br><br>Dade County | 113.2 ± 14.0;<br>range = 91–135<br>N = 11 | 140.2 ± 20.1;<br>range = 89–160<br>N = 16 |
| Kendall<br><br>Dade County | 103.6 ± 8.7;<br>range = 85–120<br>N = 22 | 122.1 ± 12.7;<br>range = 95–140<br>N = 10 |
| Lake Placid<br><br>Highlands County | 108.8 ± 9.8;<br>range = 95–130<br>N = 10 | 126.1 ± 15.3;<br>range = 100–165<br>N = 10 |

uary through September in Lake Placid at ambient temperatures as low as 15°C. In Miami, it calls from January to October (Krakauer, 1968). In Homestead, calling was heard the first week of December 1997 after 15.2 cm of rain had fallen.

Breeding sites include puddles, canals, and rock pits. Gravid females are present throughout the year in Miami (Krakauer, 1968). Egg deposition begins in March and occurs opportunistically thereafter through August (Lake Placid) or September (Homestead). Eggs are laid in strings below the water's surface. Metamorphosis can occur in less than 2 months, at body sizes of 10–13-mm SVL.

Three instances of interspecific amplexus have been observed between male *B. terrestris* and female *B. marinus* in Lake Placid.

**Diet:** *Bufo marinus* has a broad diet. Beetles (Coleoptera), earwigs (Dermaptera), and ants (Hymenoptera) comprised the bulk of its diet in Miami, Dade County (Krakauer, 1968). In that study, no overlap in diet was detected for *B. marinus* and *B. terrestris;* however, the two species compared were sampled from different locations. In areas where both species occur, diet of *B. marinus* subsumed and was broader than that of *B. terrestris* (W.E. Meshaka, Jr. and R. Powell, Unpubl. data). *Bufo marinus* is cannibalistic and also eats other anurans including *B. terrestris, B. quercicus,* and the squirrel treefrog (*Hyla squirrella*) (Rossi, 1981; R.D. Bartlett, pers. comm., 1997; W.E. Meshaka, Jr. and R. Powell, Unpubl. data). We have also found the ringneck snake (*Diadophis punctatus*) and the eastern ribbon snake (*Thamnophis sauritus*) in stomachs of and the Brahminy blindsnake (*Ramphotyphlops braminus*) in the feces of *B. marinus.*

Field tests on *B. marinus* demonstrated use of olfactory cues that facilitated location of stationary odoriferous food (Rossi, 1983). Carrion, feces, and dog food, among other stationary food items, were reported from toads collected in Tampa; however, beetles and ants were the predominant foods (Rossi, 1983).

**Predators:** The American crow (*Corvus brachyrhynchos*), red-shouldered hawk (*Buteo lineatus*), blue jay (*Cyanocitta cristata*), northern mockingbird (*Mimus polyglottos*), eastern hognose snake (*Heterodon platirhinos*), southern water snake (*Nerodia fasciata*), *Thamnophis sauritus,* common garter snake (*T. sirtalis*), and eastern indigo snake (*Drymarchon corais*) prey on *B. marinus* (Rossi, 1981; Meshaka, 1994a). The first two predators in this list are known to roll the toad onto its back and eviscerate it, thereby avoiding the potentially deadly effects of bufotoxin ingestion. The native yellow bullhead catfish (*Ictalurus natalis*) preys on the tadpoles of *B. marinus,* but the introduced walking catfish (*Clarias batrachus*) does not (Rossi, 1981).

When threatened *B. marinus* can hop away quickly. A cornered individual faces its attacker, inflates its body with air to appear larger, and tilts its head downward so that it can meet a head-on attack with toxin-filled parotoid glands. Occasionally, a threatened individual will actually lunge at an attacker.

# FAMILY LEPTODACTYLIDAE

## *Eleutherodactylus coqui* Thomas, 1966

**Common name:** Puerto Rican Coqui

**Other common name:** none

**Description:** Conant and Collins (1998) describe this species as "a small, brown or grayish brown frog with a bewildering variety of patterns." The dorsal color is highly variable, sometimes uniform gray or brown without pattern, and sometimes marked with cream dorsolateral stripes and/or dark brown marks and black spots. The venter is white or pale yellow with small brown spots. The toe pads are enlarged, and there is no webbing between the toes.

**Body size:** Florida individuals measured by Bartlett and Bartlett (1999) range 31.8–44.5-mm SVL. This species is known to reach 58-mm SVL (Conant and Collins, 1998).

*Eleutherodactylus coqui* Puerto Rican Coqui. Photo by R.D. Bartlett.

**Similar species:** This species could possibly be confused with the squirrel treefrog (*Hyla squirella*), a highly variable native species. Refer to field guides such as Bartlett and Bartlett (1999) and Conant and Collins (1998) for photographs and descriptions of this species. The only similar exotic is the greenhouse frog (*E. planirostris*) which is smaller in size, more slender in appearance, and usually has rusty overtones (Bartlett and Bartlett, 1999). The greenhouse frog also has reddish eyes. Note also that *E. coqui* has no webbing between its toes.

**History of introduction and current distribution:** *Eleutherodactylus coqui* is native to Puerto Rico. It was first recorded at Fairchild Tropical Garden in Miami, Dade County, by Austin and Schwartz (1975); however, this population became extinct during a freeze in the late 1970s (Wilson and Porras, 1983). This species is present in bromeliad nurseries in Homestead, Dade County, apparently unable to survive beyond the protective environment of greenhouses (Loftus and Herndon, 1984; W. Loftus, pers. comm., 1998). We heard *E. coqui* calling from these same greenhouses in 1999 and 2000.

*Eleutherodactylus coqui*

**Habitat and habits:** *Eleutherodactylus coqui* is nocturnal but is sometimes active on rainy days. In Florida, the species is apparently restricted to greenhouses. Although it may venture out into the areas surrounding greenhouses during the warmer months, *E. coqui* is apparently unable to survive there during the colder southern Florida winters.

**Reproduction:** In 1996, *E. coqui* was heard calling nightly between May and October from a bromeliad nursery in Homestead (O.L. Bass, Jr., pers. comm., 1998). The call is a high-pitched, two-syllable chirp sounding like "co qui," hence the common and scientific names. We have no other information regarding reproduction in Florida, but this species is known to lay its eggs on land and there is no free-swimming tadpole stage. Hatchlings appear as tiny frogs with a vestigial tail.

**Diet:** Although food habits of Florida specimens have not been studied, this species is known to eat small invertebrates, mainly insects and spiders (Kraus et al., 1999).

**Predators:** We know of no reported instances of predation on this species in Florida. Its presence inside greenhouses probably limits contact with most predators, but it is likely eaten by the Cuban treefrog (*Osteopilus septentrionalis*), a predator of frogs (Meshaka, 2001) that also inhabits greenhouses.

## *Eleutherodactylus planirostris* (Cope, 1862)

**Common name:** Greenhouse Frog

**Other common name:** none

**Description:** The dorsal body color is reddish brown with either a mottled or striped pattern. The venter is off-white to gray. The skin is slightly warty in appearance. This frog has small toe pads and reduced webbing between the toes.

**Body size:** The largest male (17.5-mm SVL) was from Miami, Dade County. The largest female (26.5-mm SVL) was from Cudjoe Key, Monroe County (Duellman and Schwartz, 1958).

**Similar species:** This species could be mistaken for a number of native species. The chorus frogs (genus *Pseudacris*) live in areas that are generally wetter than

*Eleutherodactylus planirostris* Greenhouse Frog. Photo by R.D. Bartlett.

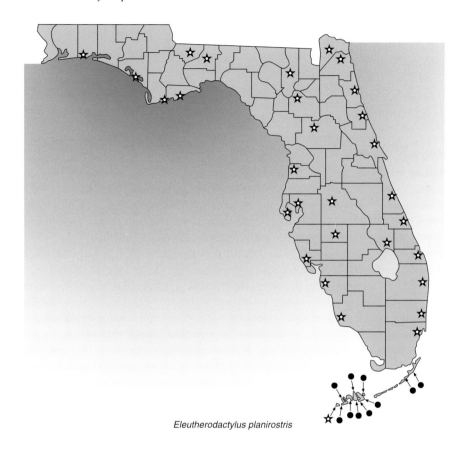

*Eleutherodactylus planirostris*

that favored by *E. planirostris* and they usually have a pattern of lines made up of spots or stripes. The spring peeper (*Pseudacris crucifer*) usually has an X-shaped mark on the back. The cricket frogs (genus *Acris*) have stripes on the posterior thigh and highly webbed feet. The squirrel treefrog (*Hyla squirella*) lacks the reddish color of *E. planirostris*. *Eleutherodactylus planirostris* differs from the exotic Puerto Rican coqui, (*E. coqui*) in that *E. planirostris* is smaller in size, more slender in appearance, has a small amount of webbing between the toes, and has an overall reddish coloration and reddish eyes.

**History of introduction and current distribution:** *Eleutherodactylus planirostris* is native to Cuba and the Bahamas. It has been known from southern Florida (Cope, 1875) and Key West, Monroe County (Cope, 1889), since the 19th century. Barbour (1910) later found it in Brevard County. Skermer (1939) reported this species in Hillsborough County, Carr (1940) reported it from Dade County, and Goin (1944) recorded it from Jacksonville, Duval County. Reichard and Stevenson (1964) reported it from Tallahassee, Leon County, in the panhandle. Goin (1947) considered populations in northern Florida to be disjunct from a continuous southern Florida range. Schwartz (1974) reported a continuous range from Leon, Duval, and Alachua counties south to Key West. Schwartz specifically listed records from Alachua, Brevard, Broward, Collier, Dade, Flagler, Indian River, Lee, Marion, Monroe, Okeechobee, Palm Beach, Pinellas, Polk, Sarasota, and Volusia counties.

The present distribution of *E. planirostris* is practically continuous statewide (Duellman and Schwartz, 1958; Schwartz, 1974; Lazell, 1989; Conant and Collins, 1998). However, its distribution is somewhat spotty in the panhandle. Recent county records include: Bay, Columbia (Conant and Collins, 1991), Martin (Timmerman et al. 1994), Okaloosa (Jensen and Palis, 1995), Gadsden (Enge, 1998), Franklin (Irwin, 1999; Krysko and Reppas, 1999), Nassau (Wray and Owen, 1999), Hernando (Enge and Wood, 1999–2000), Hardee (Christman et al., 2000), and St. Johns (Krysko and King, 2000). Localities in the Florida Keys are: Key West, Sugarloaf Key, Stock Island, Cudjoe Key, Summerland Key, Middle Torch Key, Little Torch Key, Big Pine Key, No Name Key, Upper Matecumbe Key, and Key Largo.

*Eleutherodactylus planirostris* may have dispersed to the lower Florida Keys naturally from the West Indies as well as incidentally in cargo ships. However, its rapid mainland dispersal northward has probably been facilitated by incidental transport with nursery plants (Goin, 1947). Geographic range expansion incidentally with plants would undoubtedly be faster than by natural processes and would explain the discontinuous pattern to its geographic expansion in Florida.

**Habitat and habits:** The habitat of *E. planirostris* is primarily mesophytic forests, but also prairie, pineland, mangrove forest, and disturbed sites (Duellman and Schwartz, 1958; Dalrymple, 1988). In urban areas, it often occupies potted plants and mulch piles. *Eleutherodactylus planirostris* occurs in "humid habitat that provides cover," such as woodpiles, trash piles, and leaf mold (Goin, 1947). Other authors have found it under debris (Van Hyning, 1933), under fallen logs in tropical hardwood hammocks (Carr, 1940), under heaps of limestone that trap moisture in pinelands (Deckert, 1921), and under the bark of upright trees (Harper, 1935). We have also found it under boards, rocks, and the fallen fronds of royal palm trees and coconut palm trees. This species has been taken along the ecotone of a mesic pine flatwoods and shrub-dominated dome swamp (Jensen and Palis, 1995)

*Eleutherodactylus planirostris* often shares microhabitats with other amphibians and reptiles. In sandy upland habitats of Florida, it can be the predominant vertebrate commensal of gopher tortoise (*Gopherus polyphemus*) burrows (Lips and Layne, 1989), a striking departure from mesic settings with which it is conventionally associated. We have collected *E. planirostris* and the introduced Brahminy blind snake (*Ramphotyphlops braminus*) from beneath the same cover objects in a Brazilian pepper thicket. Like Carr (1940), we have found it under the same cover as the eastern narrowmouth toad (*Gastrophryne carolinensis*).

*Eleutherodactylus planirostris* is primarily terrestrial or semifossorial. Less frequently, it has been observed above ground. At night, individuals have been observed up to 1.5 m above the ground on gumbo limbo trees and the sides of buildings in Everglades National Park. Lee (1969) found individuals under hanging sabal palm fronds, and Neill (1951) found this species in bromeliads. It is primarily nocturnal but will come out of its retreats on rainy days.

Overland movements in central Florida were greatest during fall and winter (W. Meshaka and G.E. Woolfenden, unpublished data). It was found only in the months of September through January in a swimming pool in Lake Placid, Highlands County (Table 2).

**Reproduction:** Males (15.0–17.5-mm SVL) and females (19.5–25.0-mm SVL) mature in 1 year (Goin, 1947; Duellman and Schwartz, 1958). In Gainesville, Alachua County, breeding occurs from late May to late September with a peak in July (Goin, 1947). In Everglades National Park, calling occurs from March through October. In Homestead, Dade County, calling occurs from late February through mid-November.

**Table 2.** The total number *Eleutherodactylus planirostris* captured each month in a swimming pool in Lake Placid, Highlands County, during 1991–1994.

| Month | Number of Captures |
|---|---|
| January | 1 |
| February | 0 |
| March | 0 |
| April | 0 |
| May | 0 |
| June | 0 |
| July | 0 |
| August | 0 |
| September | 1 |
| October | 5 |
| November | 8 |
| December | 1 |
| | Total: 16 |

During the breeding season, choruses that sound like soft chirping are almost nightly events and intensify with rain. Choruses in Homestead have been heard in ambient temperatures as low as 20°C. Rain, garden sprinklers, and sultry days initiate diurnal choruses.

Amplexus is axillary. Up to 26 eggs are laid in moist refuges that are attended by the male (Goin, 1947). The tadpole stage is passed in the egg, and neonates (4.3–5.7-mm SVL) hatch in 13 to 21 days and appear as diminutive versions of the adult (Goin, 1947). In Gainesville, hatchlings appear by mid-June (Goin, 1947). Lazell (1989) reported neonates from Key West near the end of May and in early June.

Reproduction in this species appears to have been influenced by Hurricane Andrew in August 1992. Preceding and during the hurricane, calling took place from flowerpots inside a house in Miami. Calling persisted for weeks after the storm during daylight hours and at dusk from inside outdoor drains (Meshaka, 1993).

**Diet:** In order of prominence, *E. planirostris* preys on ants (Hymenoptera), beetles (Coleoptera), and roaches (Dictyoptera), but it also feeds on other small invertebrates (Goin, 1947; Duellman and Schwartz, 1958).

**Predators:** Predators of *E. planirostris* include the Cuban treefrog (*Osteopilus septentrionalis*) (Meshaka, 2001) and the ringneck snake (*Diadophis punctatus*).

# FAMILY HYLIDAE

*Osteopilus septentrionalis* (Duméril and Bibron, 1841)

**Common name:** Cuban Treefrog

**Other common name:** Rana Platanera

**Description:** Subadults and adults range from faded gray to brown with chocolate brown markings to uniformly green. Metamorphs are greenish gray, often with a creamy white lateral stripe (Meshaka, 2001). The iris is gold in adults and red in metamorphs. The skin and bone of the skull are co-ossified in individuals larger than 30–35-mm SVL. Because of this skin-bone fusion, the skin on the skull has a somewhat rough texture. The toe pads are very large, about the size of the tympanum. Tadpoles are light or dark golden brown in body color. The eyes are laterally placed, the vent tube is dextral, and the spiracle is sinistral. The tail is pointed and about twice the length of the body. These tadpoles have two upper and four lower rows of teeth (the second upper and first lower are interrupted medially). They are generally found in schools.

*Osteopilus septentrionalis* Cuban Treefrog. Photo by S.L. Collins and J.T. Collins.

**Body size:** The largest male (85-mm SVL) and female (165-mm SVL) were from Lake Placid, Highlands County (Meshaka, 1996b).

**Similar species:** This frog could possibly be mistaken for one of the native treefrogs (genus *Hyla*). In proportion, no other frog in Florida (native or exotic) has such large toepads as the Cuban treefrog. The native species most often confused is the squirrel treefrog (*H. squirella*). Small *O. septentrionalis* differ from similar sized *H. squirella* in that *O. septentrionalis* has slight but noticible webbing between the fingers while *H. squirella* does not.

**History of introduction and current distribution:** *Osteopilus septentrionalis* is native to Cuba, the Cayman Islands, and the Bahamas. The first record of this treefrog in Florida was from Key West, Monroe County. Cargo ships from Cuba were suspected as the source of introduction to the Keys (Barbour, 1931). However, other researchers well acquainted with the species speculated that *O. septentrionalis* is potentially a natural disperser over saltwater (Duellman and Crombie, 1970) and might be native to the lower Florida Keys (Carr, 1940; Duellman and Schwartz, 1958; Duellman and Crombie, 1970; Lazell, 1989; Meshaka, 2001).

Until 1945, *O. septentrionalis* was probably absent from mainland Florida (A. Jones, pers. comm., 1997; W.B. Robertson, Jr., pers. comm., 1997; G. Simmons, pers. comm., 1997). However, 20 years after Barbour's (1931) record, it was detected on southern mainland Florida in Dade County (Schwartz, 1952). Soon af-

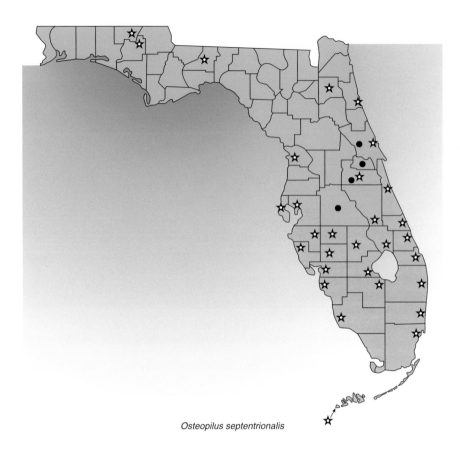

*Osteopilus septentrionalis*

ter its detection in Miami, Dade County, its dispersal through Florida was rapid, discontinuous, and associated with human activities (Meshaka, 1996a, 1996c).

This treefrog has been recorded in the following counties: Monroe (Barbour, 1931), Dade (Schwartz, 1952), Broward (King, 1960), Collier (Duellman and Crombie, 1970), Palm Beach (Austin, 1975), DeSoto, Orange (Ashton, 1976; Stevenson, 1976), Leon, Sarasota, Washington (Ashton, 1976), Highlands (Stevenson, 1976), Charlotte (Layne et al., 1977), Indian River, St. Lucie (Myers, 1977), Lee (Wilson and Porras, 1983), Pinellas (Somma and Crawford, 1993), Brevard, Citrus, Clay, Hardee, Hendry, Hillsborough, Holmes, Manatee, Martin, Osceola, Okeechobee (Meshaka, 1996a), Glades (Conant and Collins, 1998), St. Johns (Krysko and King, 1999), and Volusia (Campbell, 1999). The FLMNH has specimens from Polk and Seminole counties.

**Habitat and habits:** *Osteopilus septentrionalis* occupies many types of disturbed habitats (Carr, 1940; Duellman and Schwartz, 1958; Dalrymple, 1988; Meshaka, 2001), and it thrives around buildings. Densities on buildings can be high. For example, 105 individuals were counted at one time on a single building in Everglades National Park (Meshaka, 2001).

Characteristics of buildings most occupied by *O. septentrionalis* are large building size, the presence of outdoor lights that attract prey, the availability of retreats (e.g., the spaces behind drain pipes and air conditioners), and nearby standing water that provides breeding sites. Buildings also radiate heat, providing protection from the cold (Meshaka, 2001).

In natural areas this species is most successful in hardwood hammocks, where it is limited by the number of refuges and by low daytime relative humidity during the dry season (Meshaka, 2001). *Osteopilus septentrionalis* also inhabits mangrove forests, stands of Brazilian pepper, and pinelands that are not frequently burned (Meshaka, 2001).

This frog becomes active shortly after sunset and remains active until first light. It is active under a wide range of physical conditions, but it avoids dry, windy, moonlit nights. Tadpoles are active in the low light conditions of dawn, dusk, night, and overcast days.

This frog is active in Everglades National Park throughout the year. Peak activity is during the wet season of May to October. Activity is greatest on days when average physical conditions are warm (23.7°C), humid (94.5% RH), and wet (1.2-cm precipitation) (Meshaka, 2001). In Florida and in the West Indies, *O. septentrionalis* is a paratenic host to the nematode, *Skrjabinoptera scelopori* (Meshaka, 1996d).

**Reproduction:** Sexual maturity in *O. septentrionalis* is reached in a few months in males (27-mm SVL) and within 1 year in females (45-mm SVL). In the Everglades, average body size at sexual maturity is larger in females (65-mm SVL) than in males (45-mm SVL) (Meshaka, 2001).

The gonadal cycle is dictated by day length. Reproduction is continuous but peaks twice within the wet season of May to October in southern Florida (Meshaka, 2001). On Key West, the population breeds within the strongly bimodal wet season of the summer months (B. Ford and F. Ford, pers. comm., 1994). At the latitude of the north shore of Lake Okeechobee, gravid females have been collected throughout the year. In Tampa, cold winter temperatures decrease the likelihood of continuous reproduction (Meshaka, 2001).

Breeding sites are warm pools of water devoid of fish. Hard rains and hurricanes provide breeding sites and incite enormous breeding aggregations with deafening choruses (Meshaka, 1993, 2001). The call sounds like a rasping snore (Carr, 1940) or grating squawk (Meshaka, 2001). Males that are either warming up or calling under less than optimal conditions end their call with a series of 3–8 clicks.

Amplexus is axillary, and large males typically mate with large females (Meshaka, 2001). Males also have been observed in amplexus with green treefrogs (*Hyla cinerea*), dead southern toads (*Bufo terrestris*), and conspecific males (Meshaka, 1996e). Clutches average just over 3000 eggs, and both clutch size and ovum size increase with the body size of the female. Eggs are laid after a rain in partial clutches of 75 to 1000 eggs in a very sticky single-layer, surface film.

Eggs hatch in less than 30 hours. In water exceeding 35°C, the larval period is 3–4 weeks, but the period will surpass 3 months if the water is less than 30°C. Transformation occurs at 12–20-mm SVL. Tadpoles generally eat algae but are also cannibalistic (Crump, 1986; Babbitt and Meshaka, 2000).

**Diet:** This treefrog preys on practically anything that moves and can be swallowed in one piece. Roaches (Dictyoptera) are the main prey item consumed by *O. septentrionalis* in natural systems and on buildings (Meshaka, 2001). It also feeds on lizards (anoles and geckos), frogs, toads, and other *O. septentrionalis* (Meshaka, 2001).

*Osteopilus septentrionalis* generally captures its prey by ambush from above. It uses its large front feet to restrain and force large, struggling prey into its mouth. By this feeding method, a 65-mm SVL *O. septentrionalis* preyed on a 60-mm SVL gecko, and a 60-mm SVL *O. septentrionalis* preyed on a 40-mm TL blattid roach (Meshaka, 2001).

To avoid the noxious secretions of chemically protected prey, such as the Florida stinking roach (*Eurycotus floridana*), it systematically drains the toxins from the prey. First, it walks backwards, dragging the prey along a substrate to drain its toxins. It then brings the prey into its mouth with its forearms to taste for toxins. If toxins remain, the process is repeated (Meshaka, 2001).

**Predators:** Native snakes that prey on *O. septentrionalis* include the yellow rat snake (*Elaphe obsoleta quadrivittata*) (Meshaka and Ferster, 1995), eastern racer (*Coluber constrictor*) (Meshaka and Ferster, 1995), eastern garter snake (*Thamnophis sirtalis sirtalis*) (Meshaka and Jansen, 1997), and peninsula ribbon snake (*Thamnophis sauritus sackenii*) (Love, 1995). Native birds, such as the barred owl (*Strix varia*) (Meshaka, 1996f), and American crow (*Corvus brachyrhynchos*) (Meshaka, 2001) prey on this species. *Corvus brachyrhynchos* eat not only adult *O. septentrionalis* but also wade in water through grassy depressions for tadpoles and metamorphs (Meshaka, 2001). One exotic species known to prey on *O. septentrionalis* is the knight anole (*Anolis equestris*) (R. and J. Seavey, pers. comm., 1997).

# TESTUDINES: TURTLES

## FAMILY EMYDIDAE

### *Trachemys scripta elegans* (Wied, 1838)

**Common name:** Red-Eared Slider

**Other common name:** Pond Slider

**Description:** The carapace of juveniles is green, marked with many light and dark lines. This color gradually becomes duller in adults, eventually becoming more grayish with less pronounced markings. In old males, the carapace can be uniformly gray, olive, or almost black. A bright red horizontal mark is located behind the eye, giving this turtle its common name. This mark begins to lose its brilliance with age, especially in males. The skin of males often darkens considerably, even-

*Trachemys scripta elegans* Red-Eared Slider. Photo by R.D. Bartlett.

23

tually becoming nearly uniform black. The red mark is usually obscured in these melanistic males.

**Body size:** The largest male (180-mm CL) and female (268-mm CL) were from Snapper Creek canal in Miami, Dade County.

**Similar species:** *Trachemys scripta elegans* is superficially similar in appearance to several native species of emydid turtles from which it differs in having a red mark behind the eye. We refer you to field guides such as Bartlett and Bartlett (1999), Conant and Collins (1998), or Ashton and Ashton (1991) for descriptions of native emydids. Of note is that another subspecies (*T. s. scripta*) occurs naturally in northern Florida. Where the exotic *T. s. elegans* has come into contact with these native populations, they readily interbreed, forming intergrades and introducing exotic genes into otherwise normal populations (Bartlett and Bartlett, 1999).

**History of introduction and current distribution:** *Trachemys scripta* occurs naturally from central and eastern United States, through Mexico and Central America to Colombia and Venezuela, with disjunct populations in Brazil, Uruguay, and Argentina. Thus, this species is the widest ranging nonmarine turtle in the world (Lee, 1996). Fourteen subspecies of this turtle have been named. One subspecies, *T. s. scripta*, is native to northern Florida. The subspecies *T. s. elegans* is exotic to Florida, where it became established in canals of Dade County by 1958 (Wilson and Porras, 1983). The species is also known from Orange (Bancroft et al., 1983) and Pinellas (Hutchison, 1992) counties. Ashton and Ashton (1991) report populations from Collier, Duval, and Marion counties. We have observed it in the Dade County localities of North Miami, Kendall, South Miami, Coral Gables, and Homestead, and have seen many individuals in a large solution hole on Big Pine Key, Monroe County. One record of an adult exists for Stock Island, Monroe County (Butterfield et al., 1994a) and we have removed several adult and juvenile specimens from Wekiwa Springs State Park in Orange County. The Florida Fish and Wildlife Conservation Commission (1999–2002) re-

*Trachemys scripta elegans*

ports reliable observations in Alachua County. Our observations and those of Wilson and Porras (1983) lead us to suspect that other unreported populations occur throughout Florida and are probably the result of accidental or intentional release of pets. Indeed, on their range maps, Bartlett and Bartlett (1999) show colonies in a number of additional Florida localities and they point out that this popular pet species has become established throughout the United States and in a number of other countries.

**Habits and habitat:** *Trachemys scripta elegans* occupies canals and the ponds of wayside parks. This turtle is often seen basking, sometimes stacked two or three individuals high, on emergent structure. At some park ponds, patrons frequently feed this turtle. In these ponds, *T. s. elegans* often swims toward humans who are on the shore.

**Reproduction:** A June-August 1982 collection of *T. s. elegans* from Snapper Creek canal in Kendall, Dade County, yielded four males (148 ± 32-mm CL; range = 106–180), five females (237 ± 21-mm CL; range = 210–268) and a juvenile female (160-mm CL). One female (232-mm CL) showed evidence of a recent clutch of nine eggs. Another female (241-mm CL) contained 12 eggs (37.6 ± 1.2-mm [range = 35.5–41.0] × 24.4 ± 0.57-mm [range = 23.3–25.5]) and evidence of two earlier clutches of 11 eggs each.

**Diet:** *Trachemys scripta elegans* is an omnivore, but its diet in Florida is poorly known. In other parts of its range, this species has been shown to eat mainly vegetation, with some animal material (mollusks, insects, fish) consumed as well. Juveniles tend to eat a larger proportion of animal material (Lee, 1996; Gibbons, 1990).

**Predators:** We have no records of predators of this turtle in southern Florida.

# SQUAMATA:
# LACERTILIA—LIZARDS

## FAMILY AGAMIDAE

*Agama agama* (Linnaeus, 1758)

**Common name:** Common Agama

**Other common name:** Red-Headed Agama

**Description:** The body color of males is bluish or black. The body color of females is a drab brown. Males have a bright yellow or orange head. This species
has rough scales.

*Agama agama* Common Agama. Photo by R.D. Bartlett.

**Body size:** The largest male (115-mm SVL) and female (107-mm SVL) are both from Homestead, Dade County.

**Similar species:** It is unlikely that this species will be mistaken for any other in Florida. The most similar species, the Indochinese bloodsucker (*Calotes mystaceus*) is more slender and elongated in body form.

**History of introduction and current distribution:** *Agama agama* is native to East Africa. Two colonies of this lizard have existed in Davie, Broward County, since the mid-1980s (Bartlett and Bartlett, 1999). Both colonies are the result of releases from pet dealerships. One colony is small and in the vicinity of warehouses. The second colony adjoins a residential area and occupies about six city blocks. A small colony exists in Dade County as well (Bartlett and Bartlett, 1999). A colony located in Homestead originated from the escape and dispersal of individuals from a now defunct exotic pet dealer.

*Agama agama*

**Habitat and habits:** *Agama agama* occupies rock piles, buildings, and trees in disturbed areas and is unknown in natural areas. Individuals are most active on hot sunny days and have been seen running across roads. This species is very wary and flees when approached by humans.

**Reproduction:** *Agama agama* is an egg layer. Very little is known regarding reproduction of this species in Florida. The left testis of a male (115-mm SVL) from Homestead captured in October measured 9.8 × 7.3-mm. In their native range, females lay 5–7 eggs in a hole dug into moist, sandy soil exposed to sunlight. The eggs hatch in 8–10 weeks and the hatchlings are about 38-mm SVL (Crews et al; 1983).

**Diet:** *Agama agama* is primarily carnivorous. The brown anole (*Anolis sagrei*) appears to be uncommon at the sites occupied by this lizard, suggesting that *A. agama* may be a predator of this anole. In its native range, this species is primarily an insectivore, eating mainly ants (Hymenoptera), grasshoppers (Orthoptera), beetles (Coleoptera), and termites (Isoptera); however, it is also known to eat small mammals, small reptiles, and vegetation such as flowers, grasses, and fruits (Harris 1964).

**Predators:** We know of no reports of predation in Florida.

## *Calotes mystaceus* Duméril and Bibron, 1837

**Common name:** Indochinese Bloodsucker

**Other common names:** Bloodsucker, Tree Agama

**Description:** Males tend to be grayish in coloration with a bluish head, whereas females tend to be brown with darker longitudinal stripes and transverse bands (Bartlett and Bartlett, 1999). These lizards are large-scaled and have a serrated dorsal crest.

**Body size:** Adults reach 381-mm TL (Bartlett and Bartlett, 1999).

**Similar species:** It is unlikely that this species will be mistaken for any other in Florida. The most similar species, the common agama (*Agama agama*) is not as slender and elongated in body form as *C. mystaceus*.

**History of introduction and current distribution:** *Calotes mystaceus* is native to the East Indies. A colony has existed in Okeechobee, Okeechobee County, since the early 1980s (Bartlett and Bartlett, 1999). The source of its introduction appears

*Calotes mystaceus* Indochinese Bloodsucker. Photo by R.D. Bartlett.

to be related to the pet trade, and individuals
from this colony are often offered for sale at
local pet shops. This species also occurs in
Glades County (Bartlett and Bartlett, 1999).

**Habitat and habits:** *Calotes mystaceus* is
arboreal, and individuals of the Okeechobee
County colony occupy a citrus grove.

**Reproduction:** This species is an egg layer;
however, we have no information on reproduction for Florida populations.

**Diet:** *Calotes mystaceus* is an insectivore;
however, we have not surveyed food habits
of Florida populations.

**Predators:** We know of no reports of predation in Florida.

*Calotes mystaceus*

# FAMILY IGUANIDAE
# SUBFAMILY IGUANINAE

## *Ctenosaura pectinata* (Wiegmenn, 1834)

**Common name:** Mexican Spinytail Iguana

**Other common names:** Black Spinytail Iguana, Spinytail Iguana

**Description:** Males are black with white or yellow blotches. Females are black
with white, yellow, or peach markings, often with a green overtone. Adults have
a black head. The young are bright green with small dorsal and lateral black mark-

ings. The tail has three rows of flat scales between each of the first five or six whorls of enlarged, spiny scales and two rows of flat scales between each of the next five or six whorls of enlarged scales (Bailey, 1928).

**Body size:** The largest male (293-mm SVL) and female (253-mm SVL) were collected from Perrine, Dade County.

**Similar species:** This species resembles the black spinytail iguana (*Ctenosaura similis*) but can be distinguished by examination of tail scalation. In *C. pectinata,* the tail has three rows of flat scales between each of the first five or six whorls of enlarged, spiny scales and two rows of flat scales between each of the next five or six whorls of enlarged scales. In *C. similis,* the tail has two or three rows of flat scales between each of the first, second, and sometimes third whorls of enlarged scales and it has two rows of flat scales between each of the next six to eight whorls of enlarged scales (Bailey, 1928). This species differs from the green iguana (*Iguana iguana*) in color (at least for adults) and that *I. iguana* has no whorls of enlarged, spiny scales on the tail.

**History of introduction and current distribution:** *Ctenosaura pectinata* is native to western Mexico. Eggert (1978) first reported this lizard from Dade

*Ctenosaura pectinata* Mexican Spinytail Iguana. Photo by S.L. Collins.

County. At the time, Eggert mistakenly
identified the species as *C.
similis;* this
misidentification was later cleared up by
Wilson and Porras (1983). We corroborate
its presence in a single colony in Perrine
bordering Cutler Bay. *Ctenosaura pectinata* has been reported but is not established
in Everglades National Park (Butterfield et
al., 1997; Meshaka et al., 2000). *Ctenosaura* species are common in the pet trade.
Releases of individuals, accidental or otherwise, best explain the source of its introduction.

**Habitats and habits:** *Ctenosaura pectinata*
occupies areas with rock or cement structures such as stone walls, abandoned houses,
rock piles, and concrete slabs. Burrows are
located under concrete slabs and in rock
piles. This lizard is most active on hot,
sunny days. On mornings during May, we
have observed up to seven individuals basking on the terracotta tile roof of an abandoned house. When approached, individuals
of this very wary species quickly retreat into
nearby burrows.

*Ctenosaura pectinata*

**Reproduction:** Gravid females have been found in early June (Wilson and Porras, 1983), and a clutch of 13 eggs was found deposited in sand (Eggert, 1978).

**Diet:** *Ctenosaura pectinata* feeds on the leaves of various types of vegetation as
well as mamey fruit (Wilson and Porras, 1983). In captivity, it also feeds on fruits,
vegetables, and meat (D. Holley, pers. comm., 1994).

**Predators:** No predators of *C. pectinata* are known in Florida.

---

## *Ctenosaura similis* (Gray, 1831)

**Common name:** Black Spinytail Iguana

**Other common names:** Black Iguana, Central American Spinytail Iguana, Gar-
robo, Spinytail Iguana, Wish Willy

*Ctenosaura similis* Black Spinytail Iguana. Photo by B.P. Butterfield.

**Description:** Adult color is variable, and may include black, brown, yellow, blue, and white. The body has black crossbands terminating on the belly (Bailey, 1928). The head is usually brown. Juveniles are black, brown, and white with a green overtone. The tail has two or three rows of flat scales between each of the first, second, and sometimes third whorls of enlarged scales and it has two rows of flat scales between each of the next six to eight whorls of enlarged scales (Bailey, 1928).

**Body size:** The largest female (245-mm SVL) was from Key Biscayne, Dade County. Adult males are larger than females.

**Similar species:** This species resembles the Mexican spinytail iguana (*Ctenosaura pectinata*) but can be distinguished by examination of tail scalation. In *C. similis*, the tail has two or three rows of flat scales between each of the first, second, and sometimes third whorls of enlarged scales and it has two rows of flat scales between each of the next six to eight whorls of enlarged scales. In *C. pectinata*, the tail has three rows of flat scales between each of the first five or six whorls of enlarged, spiny scales and two rows of flat scales between each of the next five or six whorls of enlarged scales (Bailey, 1928). This species differs from the green iguana (*Iguana iguana*) in color (at least for adults) and that *I. iguana* has no whorls of enlarged, spiny scales on the tail.

**History of introduction and current distri-
bution:** *Ctenosaura similis,* native to Mexico
and Central America, has only recently been
reported from Florida (Bartlett and Bartlett,
1999), perhaps because it has been confused
with its congener, *C. pectinata.* Yet, it is now
known from many more localities in Florida
than is *C. pectinata.* Populations exist in
Boca Grande, Lee County; Davie and Holly-
wood, Broward County; and Key Biscayne
and North Miami, Dade County. The FLMNH
also lists specimens from Charlotte and Collier counties among its holdings. *Ctenosaura*
species are common in the pet trade. Releases
of individuals, accidental or otherwise, best
explain the source of its introduction.

*Ctenosaura similis*

**Habitats and habits:** *Ctenosaura similis* occupies open sunny areas near rock piles. Burrows can be found under objects such as
rocks, concrete slabs, and logs. We have also
located burrows dug directly into mounds of
dirt at an old construction site.

This iguana is most conspicuous on sunny
days, usually basking on structures such as
rock piles and stone walls. All size-classes
are frequently seen actively foraging among
bushes. A foraging lizard will typically move
a few meters and then stop for a short period
of time before continuing.

When alarmed, individuals will usually retreat into burrows. However, juveniles
will also climb into bushes to escape predators. Adult males sometimes stand their
ground, laterally compress their bodies, erect their crests, and turn their bodies
sideways toward the predator while gaping widely and hissing loudly.

**Reproduction:** Hatchlings appear in the late summer to early fall (R. St. Pierre,
pers. comm., 1995).

**Diet:** Vegetation presumably composes the bulk of the diet of this lizard, and we
have observed the seeds of Brazilian pepper in its feces. This iguana will also feed
on human garbage: on one occasion we observed an adult female feeding on a
piece of meat from a discarded roast beef sandwich.

**Predators:** No predators of *C. similis* are known in Florida.

# *Iguana iguana* (Linnaeus, 1758)

**Common name:** Green Iguana

**Other common names:** Iguana, Bamboo Chicken

**Description:** The body color is green, brownish green, or grayish green. The tail is marked with black bands. Adult males often have orange heads, legs, and sides, especially vivid in the winter. A tall dorsal crest runs from the base of the skull onto the tail. The neck region has a conspicuous dewlap. Hatchlings and juveniles are vibrant green.

**Body size:** The largest female (353-mm SVL) was from Miami, Dade County. The largest male (463-mm SVL) was from Key Biscayne, Dade County.

**Similar species:** This species superficially resembles the two introduced species of *Ctenosaura* but can be distinguished by color and examination of tail

*Iguana iguana* Green Iguana. Photo by R.D. Bartlett.

scalation. *Ctenosaura* sp. have whorls of enlarged, spiny scales on the tail which are absent in *I. iguana.* We have occasionally encountered people who have mistaken the
knight anole (*Anolis equestris*) for this
iguana. The two species share a common
green coloration (especially for smaller
specimens of *I. iguana*), but little else. The
dewlap of *A. equestris* is pink.

**History of introduction and current dis-
tribution:** *Iguana iguana* is native to Mexico, Central America, tropical South America, and the Lesser Antilles. It was first
reported from Florida in the 1960s (King
and Krakauer, 1966); however, Wilson and
Porras (1983) were the first to report breeding populations in Dade County. A breeding
colony in Collier County (Bartlett, 1980) is
still in existence (R.D. Bartlett, pers. comm.,
1998). We are aware of breeding colonies on
the lower Florida Keys, Monroe County, and
on Virginia Key and Key Biscayne, Dade
County. In extreme southern mainland
Florida, a breeding colony exists in Homestead, Dade County. Some of the colonies
were the result of release from zoological
parks and exhibits (Wilson and Porras,
1983) or of pets. Duquesnel (1998) reports

*Iguana iguana*

this species from Key Largo, Monroe County. *Iguana iguana* has been observed
in Everglades National Park but is not established (Butterfield et al., 1997; Meshaka et al., 2000). Likewise, a Lake Placid (Highlands County) report does not
represent an established colony. We have also observed this species in Broward
and Palm Beach counties and the FLMNH lists specimens from Alachua,
Broward, and Lee counties among its holdings.

**Habitats and habits:** *Iguana iguana* generally occurs in habitats containing both
trees and water. Individuals use the ground and trees for basking and foraging, and
they sleep in tree branches that hang over water. Where this species has been collected, individuals are extremely wary of humans and, if disturbed, will ascend a
tree or dive into water.

   *Iguana iguana* is active throughout the year but is sluggish on cool mornings.
Individuals are intolerant of frost. After a December 1996 heavy frost in Homestead there were several reports of dead *I. iguana* lying on the ground and

hanging in trees. However, because water retains heat, this lizard's habit of sleeping on vegetation above or near watercourses may protect it from light frost.

**Reproduction:** *Iguana iguana* mates in the winter and deposits eggs in the spring. Captives in outdoor pens in Homestead court from late December to mid-March. On 21 March 1994, a female (353-mm SVL) collected the previous day in North Miami, Dade County, deposited a clutch of 49 eggs (D. Holley, pers. comm., 1994). Ten of the eggs measured on 12 April 1994 were $43.0 \pm 0.9$-mm $\times$ $28.7 \pm 1.4$-mm. A female from Homestead, Dade County, laid a clutch of 17 eggs in April 2000. Hatchlings are common in late August (R. St. Pierre, pers. comm., 1995).

**Diet:** This lizard is primarily an herbivore. Fecal samples from a male in Homestead contained flowers, leaves, jasmine fruit, and Washington palm tree berries.

**Predators:** In Florida, domestic dogs (*Canis familiaris*) are known to kill *I. iguana*. No natural predators in Florida are known.

---

# SUBFAMILY PHRYNOSOMATINAE

## *Phrynosoma cornutum* (Harlan, 1825)

**Common name:** Texas Horned Lizard

**Other common name:** Horny Toad

**Description:** The dorsal body color is grayish brown, reddish, or buff colored with a series of darker spots along each side. The ventral surface is white. The body is dorsoventrally flattened. There are four pairs of spines along the back edge of the head, two of which are enlarged. The skin has a rough texture.

**Body size:** We have observed only road-killed (completely flattened and dried) specimens from Florida. Males of the species are known to reach 60–100-mm SVL, whereas females reach 70–120-mm SVL. Mean body size is smaller in male ($70.9 \pm 6.5$-mm SVL; range = 61.6–78.8; N = 6) than in female ($83.7 \pm 10.1$-mm SVL; range = 69.1–97.0; N = 8) museum specimens from northern Florida.

**Similar species:** This species is unlikely to be mistaken for any other in Florida.

*Phrynosoma cornutum* Texas Horned Lizard. Photo by S.L. Collins.

**History of introduction and current distribution:** *Phrynosoma cornutum* is native to the south-central United States and northern Mexico (Conant and Collins, 1998). During the early and mid-1900s, individuals were often collected and sold to tourists visiting the south-central United States. Some of these pets were then brought to Florida and subsequently released by their owners. Escaped and released pets were then reported in the scientific literature from areas throughout Florida. Therefore, many records exist where no known colonies exist. These records include the counties of Dade (De Sola, 1934), Lake (Goff, 1935), Orange, Putnam (Carr, 1940), Indian River, Marion (Allen and Neill, 1955), and Palm Beach (King and Krakauer, 1966). Allen and Neill (1955) reported rumors of this species in Polk County. The FLMNH lists a specimen from Polk County among its holdings.

A breeding population of *P. cornutum* existed in 1953 in Duval County (King and Krakauer, 1966). Interviews with individuals from Fort George Island, Duval County confirmed the continued existence of the colony in 1997. A road kill was recovered nearby, in Hugenott Park (R.D. Franz, pers. comm., 2002) as relayed by Pat Rider of White Oak Conservation Center. *Phrynosoma cornutum* was first reported from Santa Rosa Island, Escambia County, in 1940 (Carr, 1940). Although no contemporary reports of it from Santa Rosa Island were known in 1955 (Allen and Neill, 1955), a colony was recently reported at Santa Rosa Island by Jensen (1994). The FLMNH also lists specimens from Alachua, Okaloosa, and Santa Rosa counties among its holdings.

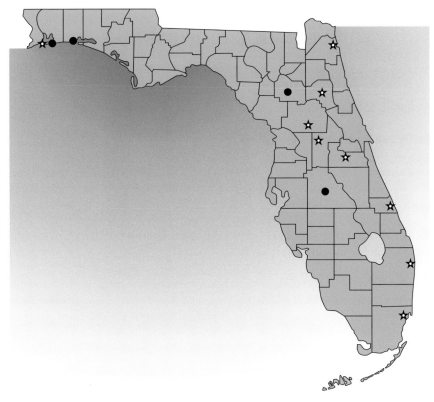

*Phrynosoma cornutum*

**Habitats and habits:** Florida populations occur in sandy, coastal areas. The Duval County population is in a natural area, but the population in Escambia County is located within a residential subdivision. Residents near the Duval County colony informed us that it was most often seen from June through August.

**Reproduction:** This species is an egg layer and has been reported to lay more than 20 eggs per clutch (Pianka, 1986); however, we have no information on reproduction for Florida populations.

**Diet:** This lizard is an ant (Hymenoptera) specialist; however, we have not surveyed food habits from Florida populations.

**Predators:** We have no information regarding predation for Florida populations.

# SUBFAMILY TROPIDURINAE

## *Leiocephalus carinatus armouri* Gray, 1827

**Common name:** Northern Curlytail Lizard

**Other common name:** Lion Lizard

**Description:** The dorsal color is brown or gray. The ventral color is cream or yellow. The dorsum is covered with large keeled scales. There are no femoral pores or enlarged postanal scales.

**Body size:** The largest male (112-mm SVL) and female (100-mm SVL) were collected from Boynton Beach, Palm Beach County.

**Similar species:** This species is superficially similar to the native Florida scrub lizard (*Sceloporus woodi*) which has femoral pores. *Leiocephalus carinatus armouri* does not have the red color or lateral skin fold of the red-sided curlytail lizard (*L. schreibersii*) and is more inclined to hold its tail in a curl than *L. schreibersii*.

*Leiocephalus carinatus armouri* Northern Curlytail Lizard. Photo by S.L. Collins.

**History of introduction and current distri-
bution:** *Leiocephalus carinatus armouri* is
native to the islands of the Little Bahama Bank
(Schwartz and Henderson, 1991). This lizard
was first reported in Florida from Palm Beach
County (Duellman and Schwartz, 1958). The
introduction resulted from a Palm Beach resident releasing 20 pairs of adults in 1945
(Weigl et al., 1969). The rate of geographic
range expansion on the mainland of Palm
Beach County by *L. c. armouri* was 87.3 and
76.7 hectares/year for the periods of 1968–
1975 and 1975–1981, respectively (Callahan,
1982).

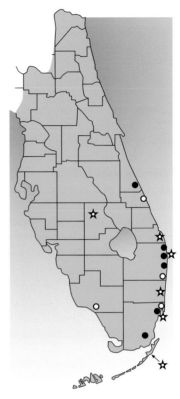

The current range of this species in
Florida includes coastal areas of Dade and
Palm Beach counties (Wilson and Porras,
1983), Broward County (Bartlett, 1994),
and Martin County (Hauge and Butterfield,
2000b). The FLMNH lists a specimen collected in Brevard County among its holdings. A population on Virginia Key was derived from the release of individuals from
zoological parks or exhibits (Wilson and
Porras, 1983). We found individuals in front
of a restaurant in Florida City, Dade County,
in March and May 2000. This colony, which

*Leiocephalus carinatus*

was still present in May 2002, is of unknown origin. One individual was observed on several occasions at Sebastian Inlet State Recreation Area, Indian
River County (B. Emanuel, pers. comm., 1999), and one individual was collected from a parking lot of a restaurant in Highlands County (Layne, 1987).
These two individuals were believed to be waifs and not part of established populations. Their presence, however, indicates that further range expansion is possible. Individuals have also been sighted at Chokoloskee, Collier County. The
species has apparently made it to the upper Florida Keys, as reported by Duquesnel (1998) who found it at John Pennekamp Coral Reef State Park on Key Largo,
Monroe County.

Two other subspecies have also been reported from Florida. *Leiocephalus c.
virescens* (Barbour, 1936) and *L. c. coryi* (King and Krakauer, 1966), both from
Dade County, represent failed colonizations.

**Habitats and habits:** *Leiocephalus carinatus armouri* occupies open, sandy and
rocky habitat. Human-disturbed areas such as rock walls, piles of rubble, and mar-
itime construction are also occupied. It excavates short burrows under rocks,

sidewalks, and other similar material that provide nighttime retreats and shelters during inclement weather.

*Leiocephalus carinatus armouri* emerges from its burrow shortly after dawn to bask on rocks or paved surfaces. Individuals also ascend trees and have been observed at heights exceeding 3 m. On 5 July 1992 at 2:00 p.m., we observed large males basking directly in the sun on open pavement. The air temperature on this occasion was approximately 33°C. During the early afternoon of 14 November 1992, we observed individuals of all size-classes basking when the air temperature was 28°C. Most of these individuals remained within 1 m of burrow entrances.

**Reproduction:** In Palm Beach, Palm Beach County, size of reproductively mature males (94.7 ± 7.0-mm SVL; range = 81.2–107.4; N = 24) averaged larger than that of reproductively mature females (82.9 ± 7.5-mm SVL; range = 70.2–94.9; N = 21). Adult males and females of Florida colonies are, on average, larger in body length than their native counterparts in the Bahamas (Callahan, 1982; Butterfield, 1996). Reasons for this size discrepancy remain unknown.

Males appeared to be most reproductively active in the summer, when their testes were enlarged (Table 3). Callahan (1982) recorded midsummer breeding and the production of a single clutch per female. Clutch size was independent of female body size. Shelled eggs were also present in females during early to midsummer (Table 4). Clutch size was small (4.0 ± 1.1 eggs/female; range = 2–6; N = 21), and shelled eggs were longer (19.2 ± 3.0-mm; range = 18.0–23.1; N = 31) than they were wide (9.9 ± 1.2-mm; range = 6.0–12.1; N = 31).

**Diet:** *Leiocephalus carinatus armouri* is primarily a sit-and-wait predator but occasionally forages actively (Callahan, 1982). Specimens collected from Palm Beach County ate invertebrate prey, especially beetles (Coleoptera), roaches (Dictyoptera), and ants (Hymenoptera) (Table 5). However, Callahan (1982) provided evidence that the depredations of *L. c. armouri* negatively impacted population densities of the brown anole (*Anolis sagrei*).

**Table 3.** Mean length and width of left testes of *Leiocephalus carinatus armouri* from Palm Beach, Palm Beach County. Length and width are expressed as a percent of SVL and followed by standard deviation.

| Month | N | Testis length | Testis width |
|-------|---|---------------|--------------|
| May | 8 | 5.9 ± 0.6 | 4.2 ± 0.4 |
| July | 12 | 5.4 ± 0.8 | 3.7 ± 0.5 |
| September | 2 | 3.6 ± 0.1 | 2.3 ± 0.4 |
| November | 2 | 3.3 ± 0.7 | 1.7 ± 0.1 |

**Table 4.** Reproductive characteristics of females of *Leiocephalus carinatus armouri* from Palm Beach, Palm Beach County.

| Month | N | Number with Oviductal Eggs | Number with Yolked Follicles | Number which were Nonreproductive |
|-------|---|-----------------------------|-------------------------------|-------------------------------------|
| May | 10 | 3 | 7 | 0 |
| July | 11 | 6 | 5 | 0 |
| August | 2 | 0 | 0 | 2 |
| September | 1 | 0 | 0 | 1 |

**Table 5.** Diet of 60 *Leiocephalus carinatus armouri* from Palm Beach, Palm Beach County. Numbers indicate number of prey (number of lizards consuming each prey category).

| Taxa of Food Items | Total |
|--------------------|-------|
| Arachnida (Aranea) | 1(1) |
| Arachnida (Acari) | 2(1) |
| Colembola | 1(1) |
| Coleoptera | 73(29) |
| Coleoptera larvae | 1(1) |
| Dermaptera | 2(2) |
| Dictyoptera | 22(18) |
| Diptera | 1(1) |
| Hemiptera | 4(1) |
| Homoptera | 1(1) |
| Hymenoptera (Formicidae) | 80(17) |
| Lepidoptera | 2(1) |
| Lepidoptera larvae | 1(1) |
| Orthoptera | 1(1) |

**Predators:** Potential predators include the feral domestic cat (*Felis catus*) and eastern racer (*Coluber constrictor*), both of which are occasionally seen in areas inhabited by *L. c. armouri*.

---

## *Leiocephalus schreibersii* (Gravenhorst, 1837)

**Common name:** Red-Sided Curlytail Lizard

**Other common name:** none

**Description:** The dorsal color is brownish gray. The sides are dark red with small lightly colored spots. There is a lateral fold of skin along each side of the body. This lizard does not have femoral pores or enlarged postanal scales.

**Body size:** No Florida specimens were available, but adults are usually less than 254-mm TL (Bartlett and Bartlett, 1999).

**Similar species:** This species is superficially similar to the native Florida scrub lizard (*Sceloporus woodi*) which has femoral pores. The red color on the sides and

*Leiocephalus schreibersii* Red-Sided Curlytail Lizard. Photo by S.L. Collins.

lateral fold of skin of this species differentiate it from the northern curlytail lizard (*L. carinatus armouri*). Also, it does not curl its tail as tightly as *L. carinatus armouri*.

**History of introduction and current distribution:** This lizard is native to Hispaniola. It has been established in Dade County since 1978 when individuals imported from the northern coast of Haiti escaped from an animal dealer. Although that population appeared to have been extirpated in 1981, a second colony in Miami Lakes was derived from the original colony (Wilson and Porras, 1983). Bartlett (1994) reported a small colony near the Miami International Airport, Dade County and one in Broward County (Bartlett and Bartlett, 1999). We are also aware of a small population in North Miami, Dade County (R. St. Pierre, pers. comm., 1995).

**Habitats and habits:** We have little information for Florida populations. The North Miami population occurs along a railroad track in a residential area (R. St. Pierre, pers. comm., 1995).

*Leiocephalus schreibersii*

**Reproduction:** This species is an egg layer; however, we have no information for Florida populations.

**Diet:** Like *L. carinatus armouri*, this species is primarily an insectivore. We have not surveyed food habits of specimens from Florida.

**Predators:** We know of no reports regarding predators for Florida populations.

---

# SUBFAMILY POLYCHROTINAE

*Anolis chlorocyanus* Duméril and Bibron, 1837

**Common name:** Hispaniolan Green Anole

**Other common names:** Haitian Green Anole, Blue-green Anole

*Anolis chlorocyanus* Hispaniolan Green Anole. Photo by R.D. Bartlett.

**Description:** Body color ranges from green to black in metachrosis. The dewlap is black and blue in males.

**Body size:** The largest male (70-mm SVL) and the largest female (55-mm SVL) were from Broward County.

**Similar species:** This species superficially resembles the native green anole (*Anolis carolinensis*) whose dewlaps range in color from gray to pink to red and the Cuban green anole (*A. porcatus*) whose dewlap is pinkish purple. The Jamaica giant anole (*A. garmani*), also green in body color, has a vertebral crest and a brownish yellow dewlap.

**History of introduction and current distribution:** *Anolis chlorocyanus* is endemic to Hispaniola. Bartlett (1988) reported a population of this species from Miami, Dade County; however, he later reported that following the construction of a train station, this population no longer existed. A second colony has existed in Broward County since 1987 (Butterfield et al., 1994b). Both populations were the result of specimens being released accidentally or intentionally from pet dealerships. The Florida Fish and Wildlife Conservation Commission (1999–2002) reports that this species was established for a time in Martin County. The population, which was established on a reptile dealer's property, died out due to freezing temperatures.

**Habits and habitats:** *Anolis chlorocyanus* occupies the outside walls of buildings and the trunk-crown region of trees. Generally, it is found at 2 or more meters above the ground in trees or on buildings.

The population density of this species may be low. Fewer than 15 individuals of *A. chlorocyanus* were typically observed per visit during surveys to the Broward County site during the 1990s. Recent changes to the habitat at this site (especially the removal of many of the trees) may negatively impact this population. Habitat changes, limited geographic range, and small population sizes may jeopardize the persistence of *A. chlorocyanus* in Florida.

**Reproduction:** A female (55-mm SVL), collected 14 September 1991, contained a shelled oviductal egg (11.5 × 5.5-mm) and an enlarged follicle (5.9 × 4.1-mm). A male (70-mm SVL), collected on the same date, possessed enlarged testes (5.2 × 3.8-mm and 5.1 × 3.5-mm), and convoluted epididymides.

*Anolis chlorocyanus*

**Diet:** We have not surveyed food habits for Florida populations. Like most trunk-crown anoles, this species eats a variety of arthropods.

**Predators:** We have no information regarding predators for Florida populations.

---

# *Anolis cristatellus* Duméril and Bibron, 1837

**Common name:** Puerto Rican Crested Anole

**Other common name:** Crested Anole

**Description:** Body color ranges from greenish gray to dark brown in metachrosis. The dewlap is reddish orange with a central yellowish green area. Adult males often have a large dorsal crest on their bodies and tails.

*Anolis cristatellus* Puerto Rican Crested Anole.
Photo by S.L. Collins.

**Body size:** The largest male (72-mm SVL) was from South Miami, Dade County. The largest female (56-mm SVL) was from Key Biscayne, Dade County (Brach, 1977).

**Similar species:** This species most closely resembles the brown anole (*Anolis sagrei*), whose dewlap is red or reddish orange with a light (cream or yellow) border. The dewlap of another similar species, the largehead anole (*A. cybotes*), is creamy yellow.

**History of introduction and current distribution:** Schwartz and Thomas (1975) first reported *A. cristatellus*, native to Puerto Rico and the Virgin Islands, from Florida on Key Biscayne, Dade County. The colony was four blocks in size and located at the southern end of the Key (Brach, 1977). Another population at the old Crandon Park Zoo grounds was derived from the original colony. In 1976, a colony was reported from Red Road in South Miami, Dade

County (Wilson and Porras, 1983). Both of
these populations are still in existence, and
the latter populations are known to have
been deliberate introductions. Two other
populations in North Miami, Dade County,
also resulted from deliberate introductions
in the 1980s. Seigel et al. (1999) reported *A.
cristatellus* from Brevard County but indi-
cated that subsequent trips to the area failed
to turn up more specimens. The source of
that colonization attempt is unknown.

**Habitats and habits:** *Anolis cristatellus*
prefers partly shady areas near sunny locations. This anole is a trunk-ground lizard in
the use of its structural habitat. Large males
are conspicuous on tree trunks, fences, and
buildings. Males usually maintain a head-
down position on such structures, sometimes
perching upside down under horizontal
branches. Perch heights for males vary, but
most are found at the midtrunk level (Table
6). Females and juveniles are generally found
in more concealed locations near retreats and
close to the ground. Recent hatchlings are
most often found on the ground near the bases
of trees.

*Anolis cristatellus*

All known colonies of *A. cristatellus* be-
came established at sites already colonized by another trunk-ground lizard, *A.
sagrei*. These colonies have persisted and have expanded slowly, despite the presence of *A. sagrei*. Salzburg (1984) suggested that these two anoles might compete
for structural resources in southern Florida. In areas where both species occurred
together, perch heights of the male *A. sagrei* were lower than those of *A. cristatellus* (Salzburg, 1984). Our observations are similar to those of Salzburg. Behav-
ioral interactions between males of both species suggest that the larger-bodied *A.
cristatellus* socially dominates *A. sagrei*. Thus, the slow rate at which *A. cristatellus* has dispersed is probably not caused by competition for structural resources
between males of *A. cristatellus* and *A. sagrei*. Therefore, comparisons between
the two species along other resource axes should be investigated.

We have observed that *A. cristatellus* occupies shadier habitats than *A. sagrei*.
Thermal preferences may be important factors in shaping the distributions and
abundances of these two species.

**Reproduction:** *Anolis cristatellus* breeds from March through November. During
this time period females are gravid and males have enlarged testes (Table 7). Most
females lay one egg (9.6 ± 0.9-mm; range = 8.2–11.1; N = 25 × 5.2 ± 0.6-mm;

**Table 6.** Perch heights in cm of male *Anolis cristatellus* in April from Dade County.

| Perch height (cm) | Number of Males |
|-------------------|-----------------|
| 0–60 | 5 |
| 60–105 | 16 |
| 105–180 | 20 |
| 180+ | 14 |
| | Total 55 |

range = 4.26–5.4; N = 25) at a time, although a few females were found with two shelled eggs. Likewise, Brach (1977) found gravid females and sexually active males in April from Key Biscayne. Mean body size of mature males (61.4 ± 6.1-mm SVL; range = 46.0–70.7; N = 48) is larger than that of females (44.9 ± 2.4-mm SVL; range = 39.3–48.6; N = 31). On Key Biscayne, seven males averaged 68-mm SVL (range = 64–71-mm SVL) and six females averaged 48-mm SVL (range = 46–56-mm SVL) (Brach, 1977). A shelled egg, which was removed from a female on 23 April and incubated at 27°C, hatched on 10 June and the hatchling measured 17-mm SVL (Brach, 1977).

**Diet:** Males and females both feed primarily on ants (Hymenoptera) and beetles (Coleoptera) but also consume other invertebrate prey (Table 8). Brach (1977) found mostly beetles, caterpillars (Lepidoptera), halictid bees (Hymenoptera), and spiders (Aranea) in the stomachs of a small sample from Key Biscayne. On one occasion, we observed a large male defecating the remains of a

**Table 7.** Mean length and width of left testes of *Anolis cristatellus* from Dade County during 1991–1992. Length and width are expressed as a percent of SVL and followed by standard deviation.

| Date | N | Testis length | Testis width |
|------|---|---------------|--------------|
| March | 8 | 7.1 ± 0.7 | 4.8 ± 0.5 |
| May | 22 | 8.3 ± 1.1 | 5.9 ± 0.5 |
| September | 2 | 7.6 ± 0.2 | 5.0 ± 0.1 |
| November | 16 | 6.6 ± 1.6 | 4.5 ± 1.2 |

**Table 8.** Diet of 28 male and 22 female *Anolis cristatellus* from Dade County. Numbers indicate number of prey (number of lizards consuming each prey category).

| Taxa of Prey Items | Males | Females | Total |
|---|---|---|---|
| Hymenoptera (Formicidae) | 119(16) | 152(17) | 270(33) |
| Hymenoptera (Other) | 3(3) | 9(4) | 12(7) |
| Coleoptera | 21(9) | 9(6) | 30(15) |
| Coleoptera Larvae | 2(1) | 4(1) | 6(2) |
| Dermaptera | 5(1) | 9(3) | 14(4) |
| Lepidoptera | 3(1) | 2(2) | 5(3) |
| Lepidopteran Larvae | 2(2) | 2(1) | 4(3) |
| Hemiptera | 0(0) | 7(6) | 7(6) |
| Homoptera | 0(0) | 2(2) | 2(2) |
| Orthoptera | 0(0) | 1(1) | 1(1) |
| Blattoidea | 3(3) | 0(0) | 3(3) |
| Diptera | 5(3) | 6(3) | 11(6) |
| Unidentified Insect | 21(8) | 8(6) | 29(14) |
| Chilopoda | 0(0) | 1(1) | 1(1) |
| Diplopoda | 3(1) | 0(0) | 3(1) |
| Arachnida (Aranea) | 3(3) | 1(1) | 4(4) |
| Arachnida (Chernetidae) | 2(2) | 0(0) | 2(2) |
| *Ficus* Fruit | 4(2) | 0(0) | 4(2) |
| Shed skin | 1(1) | 1(1) | 2(2) |

Brahminy blind snake (*Ramphotyphlops braminus*). This species also occasionally eats blossoms and fruits of ficus trees (Brach, 1977; Bartlett and Bartlett, 1999).

**Predators:** We have no information on predators of Florida populations.

## *Anolis cybotes* Cope, 1862

**Common name:** Largehead Anole

**Other common name:** none

**Description:** Body color ranges from greenish gray to dark brown in metachrosis. The dewlap is cream-colored or creamy yellow.

**Body size:** The largest male (71.4-mm SVL) and female (56.7-mm SVL) were collected from Broward County.

**Similar species:** This species is similar to two other exotic anoles. The brown anole (*Anolis sagrei*) has a reddish orange dewlap with a yellow or cream border. The dewlap of the Puerto Rican crested anole (*A. cristatellus*) is reddish orange with a central yellowish green area.

*Anolis cybotes* Largehead Anole. Photo by R.D. Bartlett.

**History of introduction and current distribution:** *Anolis cybotes* is native to Hispaniola. This anole occurs at three locations in southern Florida. It was first reported from northern Dade County in 1973, the result of an intentional introduction (Ober, 1973). A second colony was reported from northern Broward County at a site formerly occupied by a pet dealership (Butterfield et al., 1994b). The Florida Fish and Wildlife Conservation Commission (1999–2002) reports that this species is also established in Martin County. The population, which was established in 1986 on a reptile dealer's property, was still there and had moved into the surrounding area in 2002. The FLMNH lists a specimen from Martin County among its holdings.

**Habitats and habits:** *Anolis cybotes* is a trunk-ground lizard. It occupies the trunks of ficus trees and human structures such as buildings, woodpiles, and fences. Large males are often seen in a vertical, head-down posture with the head extended parallel to the ground. Females and juveniles are often less conspicuous than males and stay near cover.

*Anolis cybotes*

*Anolis cybotes* can be locally abundant. Population density of the Broward County colony may be very high, typically more than 50 individuals were observed per visit to this site during surveys in the 1990s.

**Reproduction:** A female collected on 14 September 1991 contained an oviductal egg ($10.7 \times 6.0$-mm) and an enlarged follicle (9.3-mm). Left testis lengths of two males (71.4 and 56.7-mm SVL) collected on the same date were large ($6.2 \times 3.9$-mm and $6.1 \times 3.5$-mm, respectively). The left testis of a smaller male, 47.7-mm SVL measured $3.2 \times 2.2$-mm.

**Diet:** We have not surveyed food habits for Florida populations. However, *A. cybotes* preys on insects and geckos (*Sphaerodactylus*) in Hispaniola (Schwartz and Henderson, 1991).

**Predators:** We have no information regarding predators of Florida populations.

## *Anolis distichus* Cope, 1862

**Common name:** Bark Anole

**Other common name:** none

**Description:** The color of the Dominican bark anole, *Anolis distichus dominicensis,* ranges from yellowish green to dark brown in metachrosis. The dewlap of *A. d. dominicensis* is yellow with an orange central spot. The color of the Florida bark anole, *A. d. floridanus,* ranges from light gray to dark brown in metachrosis. The dewlap of *A. d. floridanus* is pale yellow. *Anolis d. floridanus* is smaller in body size than *A. d. dominicensis.* In Broward and Dade counties, subspecific designation of individuals is impossible because individuals may share characters of both subspecies (Miyamoto et al., 1986).

**Body size:** The largest male (53-mm SVL) was collected near the Miami International Airport, Miami, Dade County. The largest female (43-mm SVL) was collected from Florida International University, North Campus, Miami, Dade County.

**Similar species:** The small body size of this species makes it unlikely to be confused with other anole species. The only other anole in Florida with a pale yellow dewlap is the largehead anole (*Anolis cybotes*) which is much larger in size.

*Anolis distichus* Bark Anole. Photo by R.D. Bartlett.

**History of introduction and current distribution:** *Anolis distichus* is native to the Bahamas and Hispaniola. Of 18 recognized subspecies, two and possibly three are found in Florida. *Anolis d. floridanus* is one of Florida's earliest recorded colonizing species, having been first recorded from Miami, Dade County (Smith and McCauley, 1948). Wilson and Porras (1983) considered *A. d. floridanus* a member of the native Florida herpetofauna; however, Schwartz (1971) postulated that *A. d. floridanus* either differentiated locally in Florida or represented a western Andros population that had become established in Florida. *Anolis d. dominicensis* from Hispaniola was first reported from the Tamiami Canal area in Miami, Dade County (King and Krakauer, 1966). This lizard was suspected as having been introduced incidentally with cargo from Hispaniola. A third subspecies, *A. d. biminiensis,* from Bimini in the Bahamas, has been reported from Lake Worth, Palm Beach County (Bartlett, 1995a). *Anolis d. ignigularis,* reported by King and Krakauer (1966), represents a failed introduction.

Anolis distichus

In Dade County, some individuals share characteristics of both *A. d. floridanus* and *A. d. dominicensis* (Miyamoto et al., 1986; Butterfield, 1996). *Anolis d. floridanus* may well have once been a Florida native; however, the introduction of the Dominican form has greatly compromised the genetic integrity of the original Florida form. Now, the bark anole of Florida is a new phenotype, the result of two distinct subspecies mixing followed by the processes of drift and/or directional selection (Butterfield, 1996).

*Anolis distichus* has been expanding its range northward. This lizard was first observed in Coral Springs, Broward County, in 1988 (R. Kilhefner, pers. comm., 1991; Reppas et al., 1999). We found it in Boynton Beach, Palm Beach County, in 1991. We first observed *A. distichus* in Palm Beach, Palm Beach County, in 1995. Extensive searching for this anole in the same location in 1991 revealed none. We also found *A. distichus* on Ramrod Key, Monroe County in 1993 and at Hobe Sound Beach, Martin County in 1997.

Literature reports of this species include Broward County (Wilson and Porras, 1983; Reppas et al., 1999); Dade County (Wilson and Porras, 1983); Key West, Monroe County (Lazell, 1989); Lee County (Bartlett, 1994); and Key Vaca, Monroe County (Watkins-Colwell and Watkins-Colwell, 1995a). This lizard has also been reported from Everglades National Park, but it is not established (Butterfield et al., 1997; Meshaka et al., 2000). *Anolis distichus* occurs in Homestead and is somewhat continuous in its distribution along eastern Dade County.

**Habitats and habits:** *Anolis distichus* is a midtrunk lizard in the use of its structural habitat; most are found 105–180-cm above the ground (Meshaka, 1999a, 1999b). It is found on trees with smooth bark and on the exteriors of buildings. This anole is cryptically colored, making it difficult to detect. *Anolis distichus* is most often observed in the shade or in areas of dappled sunlight. Doan (1996) found that individuals of both sexes basked most frequently in filtered sunlight, but the sexes differed in basking frequency.

When disturbed by humans, it may escape upward or downward, often in an erratic fashion; or it may actually jump on the intruder. If the anole takes a descending escape route, it will usually take refuge under an object such as a piece of loose tree bark. However, *A. distichus* most often flees upward in spirals around the tree until out of view, only to reappear soon thereafter in the canopy.

*Anolis distichus* is active at a mean cloacal temperature of 30.5°C (range 21.3–33.6°C) (King, 1966). During cooler times of the year, most activity occurs during the middle portion of the day; however, during hotter months, peak activity occurs during the early morning and late afternoon (King, 1966). During periods of inactivity caused by extreme temperatures, refuge is taken among palm fronds and in tree crevices (King, 1966). Population densities can be high. We have observed more than 20 individuals on a single large ficus tree.

**Reproduction:** In Miami, Dade County, the reproductive season is from February through October (King, 1966). From a Miami, Dade County, sample, the mean body length of males (48.7 ± 1.9-mm SVL; range = 44.8–51.3; N = 17) was larger than that of females (41.5 ± 1.8-mm SVL; range = 39.0–45.0; N = 17). King (1966) reported the following reproductive information for *A. distichus*. Males are reproductive at 41-mm SVL and females become reproductive at 38-mm SVL. Shelled oviductal eggs range 10.0–12.0 mm in length and 6.0–7.0 mm in width. A single female is capable of laying an egg every 14 days and has the potential to lay 16 eggs per year. Eggs are deposited in leaf litter at the base of trees.

**Diet:** King (1966) found that ants (Hymenoptera) were the most important food item of *A. distichus;* beetles (Coleoptera) were the second most important. Other arthropods are also occasionally eaten. Ants (Hymenoptera) predominate in a small sample from South Miami (Table 9).

**Predators:** Ringneck snakes (*Diadophis punctatus*) and ants prey on the eggs of *A. distichus* (King, 1966). Cuban green anoles (*A. porcatus*) prey on *A. distichus.*

**Table 9.** Diet of five male and five female *Anolis distichus* from South Miami, Dade County. Numbers indicate number of prey (number of lizards consuming each prey category).

| Taxa of Prey Items | Male | Female | Total |
|---|---|---|---|
| Coleoptera | 1(1) | 0(0) | 1(1) |
| Hymenoptera (Formicidae) | 28(5) | 115(5) | 143(10) |
| Arachnida | 0(0) | 2(2) | 2(2) |

## *Anolis equestris* Merrem, 1820

**Common name:** Knight Anole

**Other common name:** Chipojo

**Description:** Adults are bright grassy green with a yellow shoulder stripe, the green areas becoming dark brown in metachrosis. Hatchlings and juveniles are green with cream-colored transverse bands along the body. The dewlap, present in males and females, is light pink.

*Anolis equestris* Knight Anole (adult). Photo by B. Stith.

*Anolis equestris* Knight Anole ( juvenile). Photo by R.D. Bartlett.

**Body size:** The largest male (180-mm SVL) and female (160-mm SVL) were from Coconut Grove, Dade County.

**Similar species:** The very large body size of this species makes it unlikely to be mistaken for any other anole. The very light pink (almost white) dewlap, present in both sexes, is unique among anoles in Florida. We have occasionally encountered people who mistook this species for the green iguana (*Iguana iguana*), with which it shares a common color and little else. *Iguana iguana* does not have a pink dewlap.

**History of introduction and current distribution:** Initial introduction of this Cuban endemic was to an unspecified location in southern Florida (Neill, 1957). A thriving population of *A. equestris* in Coral Gables, Dade County, originated from a deliberate release of individuals by a student at the University of Miami (King and Krakauer, 1966). Since its introduction in the mid-1950s, *A. equestris* has expanded its geographic range in Florida, mainly in the southernmost and coastal regions of southern Florida, including Elliott Key, Dade County (Brown, 1972); Plantation Key, Monroe County (Achor and Moler, 1982); and more recently from Naples, Collier County (Noonan, 1995). *Anolis equestris* was known from Key West, Monroe County, prior to Hurricane Andrew in 1992 (R. Ehrig, pers. comm., 1994). Miamians, displaced by the hurricane, intentionally released *A. equestris* in the Lower Keys (R. Ehrig, pers. comm., 1994). Since Hurricane Andrew, it has become more widespread in southern Florida and even locally abundant in Homestead, Dade County. A Highlands County report refers to an adult male waif found in an agricultural shipment sent from Miami to Lake Placid in March 1995. The FLMNH lists a single specimen collected in Broward County in 1974 among its holdings. A juvenile (77-mm SVL) was collected from a wild tamarind tree in March 2000 on Parachute Key in Everglades National Park. Earlier reports exist for *A. equestris* in the park (Butterfield et al., 1997; Meshaka et al., 2000) but it does not appear to be established. *Anolis equestris* has also been reported from the counties of Martin and Lee; however, Bartlett and Bartlett (1999) indicate that all *A. equestris* reported from Lee County are actually Jamaican giant anoles (*A. garmani*). The Florida Fish and Wildlife Conservation Commission (1999–2002) reports that the Martin County population, which was established in 1986 on a reptile dealer's property, was still there in 2002 (two specimens are listed in the FLMNH holdings). They also report four specimens found in Polk County, but do not report the species as breeding there.

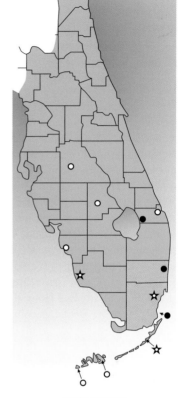

*Anolis equestris*

**Habitat and habits:** *Anolis equestris* is typically associated with the canopies of trees. The trees most often inhabited are mahogany, black olive, wild tamarind, and mango. These trees are generally closely spaced, but positioned so that some sunlight reaches their trunks.

Although *A. equestris* typically occupies the tree canopy, adults will cross phone lines or open ground to move between trees. This anole also will use the lower levels of trees for the activities of feeding, basking, and mating. The time it spends near the ground varies according to the time of the day and varies among populations (Meshaka, 1999a, 1999b; Meshaka and Rice, In Review). For example, in Homestead during August, individuals conspicuously perch within 2 m of the ground in high numbers from early to late morning. Later in the day, activity becomes restricted almost exclusively to the upper reaches of trees. Differential use of the lower levels of trees in the morning accounts for the generally higher numbers of anoles observed in the morning (Meshaka and Rice, In Review).

Use of the lower levels of trees by *A. equestris* varied among three Dade County sites. Individuals from Homestead routinely perched within 2 m of the ground

(Meshaka and Rice, In Review). *Anolis equestris* of a South Miami site (Meshaka, 1999a) were observed at similar heights, but less frequently than in Homestead. In contrast, *A. equestris* at the Coconut Grove (Meshaka, 1999b) site were not observed below 2 m, and most individuals were observed at least 3 m above the ground.

Differential perch height among sites may have been a consequence of the presence of terrestrial predators. The Coconut Grove site was characterized by the presence of possible diurnal vertebrate predators of *A. equestris*. For example, between 1–6 eastern racers (*Coluber constrictor*) were observed during each 90-minute survey at the Coconut Grove site in spring and summer. These snakes were seen hunting along vegetation edges and around the bases of trees. This snake species was seldom seen at the South Miami site (Meshaka, 1999a) and was not seen at the Homestead site. Additionally, few other potential terrestrial predators of *A. equestris* were observed at the South Miami site or the Homestead site. These observations suggest that terrestrial predator intensity might be forcing *A. equestris* to occupy the canopy. The absence or rarity of terrestrial predators in urban areas may have provided this anole with an ecological release into lower heights that they would otherwise avoid.

Activity of *A. equestris* is highly seasonal. From the end of October to the end of March in Miami and Homestead, Dade County, it is rarely seen. In April, individuals are more frequently observed, and from May through October, the lizard is seemingly everywhere. This coincides with average monthly high temperatures of $> 29°C$. In some areas during its active season, *A. equestris* can be seen on nearly every large tree in some neighborhoods.

In the middle of summer, *A. equestris* begins activity early in the morning (Meshaka and Rice, In Review). It positions itself head-down with its head angled slightly outward. By midmorning, the number of active lizards peaks and continues until late afternoon when individuals slowly become less obvious. Activity usually ends at sunset, but can extend until last light.

Diel activity in August is unimodal. Cloud cover, wind velocity, and relative humidity seem to have much less to do with activity than does ambient temperature. At temperatures below $30°C$, *A. equestris* is rarely observed (Meshaka, 1999c; Meshaka and Rice, In Review). Consequently, it is less active from November through March because the ambient temperature approaches $30°C$ less frequently as compared to other times of the year. Minimum daytime high temperature best explains the seasonal component to its activity. This observation is consistent with observations by Wilson and Porras (1983) that *A. equestris* prefers the hottest days of the year.

Juveniles, with their distinct markings, are rarely encountered. On one occasion a juvenile (97-mm SVL), whose stripes were beginning to fade, was captured near the top of a wild tamarind tree. On another occasion a striped juvenile was observed basking at 7:30 am approximately 2.5 m from the ground in a wild tamarind tree. On three different occasions, juveniles were observed less than 1 m from the ground in hedges.

Population sizes of *A. equestris* can range from 3.3/ha in a wild tamarind grove in Homestead to 29.5/ha in a tropical garden (Dalrymple, 1980). Hurricane Andrew did not adversely affect populations of *A. equestris* in Miami, Dade County, in 1992. The hurricane opened the canopy, which may have benefited this lizard by providing it with more basking sites (Meshaka, 1993). Five years after Hurricane Andrew, *A. equestris* was still seen as frequently as before the storm.

**Reproduction:** In the Kendall-South Miami area of Dade County, 24 adult males (156.4 ± 14.3-mm SVL; range = 107–174) were larger than seven females (134.6 ± 13.4-mm SVL; range = 107–150) collected in June and July. In Homestead, Dade County males (163.36 ± 9.2-mm SVL; range = 152–180; N = 9) were larger than females (137.8 ± 8.6-mm SVL; range = 123–150; N = 6). In Dade County, copulating pairs were most often observed from April through August, although a copulating pair was also observed during February in Homestead (Meshaka and Rice, In Review). A copulating pair was observed 1 m above the ground on the trunk of a mahogany tree. Three other copulating pairs were observed greater than 2.0 m above the ground on fig, wild tamarind, and sabal palm trees.

The following mating sequence was observed for one male: he first followed a receptive female from directly behind, he then caught up to her, crawled directly on top of her until his head was approximately midway up her back, and in a single motion slid his tail under hers to connect for copulation. Neither this male, nor others observed in copula, were observed biting the neck of the female, as observed in some other anoles. Mating pairs appear to face any direction except downward.

Nine shelled eggs removed from preserved females collected in June and July were longer (21.1 ± 0.91-mm; range = 18.8–22.0) than they were wide (11.4 ± 1.1-mm; range = 10.2–14.0). Mean length and width of the left testis, as a percentage of the body length, was similar in 15 males collected in June (7.3 ± 0.6%; range = 6.3–8.3 × 5.4 ± 0.5%; range = 4.5–6.2) and nine males collected in July (6.8 ± 0.7%; range = 5.3–8.3 × 5.0 ± 0.4%; range = 4.3–5.5).

In Homestead, growth to sexual maturity (100–110-mm SVL) is rapid for males (12–13 months) and females (8–9 months). Populations turn over in almost seven years but adults may live past 10 years of age (Meshaka and Rice, In Review).

**Diet:** Published accounts indicate that *A. equestris* is an omnivore that feeds on a wide range of invertebrates (Brach, 1976; Dalrymple, 1980). This lizard has also been observed to prey on the following vertebrates: Cuban treefrog (*Osteopilus septentrionalis*) (R. and J. Seavey, pers. comm., 1997), brown anole (*Anolis sagrei*) (R. Galleno, pers. comm., 1993; Meshaka and Rice, In Review), the blue-gray gnatcatcher (*Polioptila coerulea*) (Meshaka and Rice, In Review), and the Indo-Pacific gecko (*Hemidactylus garnotii*). Anolis equestris may also prey on Cuban green anoles (*A. porcatus*) (Meshaka et al., 1997a). There is evidence that *A. equestris* raids nest boxes to feed on eggs and hatchlings of purple martins (*Progne subis*) and screech owls (*Otus asio*) (J.C. Ogden, pers. comm., 1998). Individuals will also feed on vegetation, including fig berries, palm berries, and mango sap.

*Anolis equestris* has been observed to eat *O. septentrionalis* by pulling them out of bromeliads located on branches and at the bases of trees (R. and J. Seavey, pers. comm., 1997). In one instance, an adult *A. equestris* beat to death and ate a 60–70-mm SVL *O. septentrionalis* (R. Seavey, pers. comm., 1997). The anole had captured and attempted to eat the frog headfirst. The frog, vocalizing at the time, immediately inflated its body with air and kicked to obstruct the lizard's efforts. After about 10 minutes, the lizard maneuvered the frog so that it had the frog by its rear legs. With the frog hanging out of its mouth and vocalizing, the anole ascended until it found a fork in the tree. Positioning itself at the level of the fork, the anole rapped the frog in the fork approximately 20 times in about 10 seconds. The frog fell limp and became deflated. The anole then repositioned the frog headfirst in about 10 seconds and swallowed it whole in less than 1 minute.

On another occasion, a large *A. equestris* was observed to bite and hold a mourning dove (*Zenaida macroura*) by the breast (T. Chapman, pers. comm., 1998); however, the dove escaped within 1 minute. We do not know if this was a predatory attempt or if the anole mistook the mourning dove for an invader that approached too closely. In Homestead, a juvenile (ca. 60-mm SVL) from above stalked a female *A. sagrei* that was near the base of the trunk of a wild tamarind tree. In this case, the juvenile did not successfully capture the anole.

*Anolis equestris* apparently learns to avoid the southeastern lubber grasshopper (*Romalea microptera*). Young individuals grab and then instantly reject this potential prey item. We have observed adults watch *R. microptera*, but make no attempt to capture them.

**Predators:** We have not observed predation of *A. equestris* in Florida.

---

## *Anolis garmani* Stejneger, 1899

**Common name:** Jamaican Giant Anole

**Other common name:** none

**Description:** Body color ranges from bright grassy green to dark green with a gold flecked pattern, to solid black in metachrosis. The dewlap is brownish yellow. This lizard has a short spiny vertebral crest.

**Body size:** The largest male (92.4-mm SVL) and female (78.0-mm SVL) were from South Miami, Dade County.

**Similar species:** This species, at smaller body sizes, superficially most resembles the green anole (*Anolis carolinensis*), whose dewlap is highly variable in color

*Anolis garmani* Jamaican Giant Anole. Photo by R.D. Bartlett.

(gray, red, pink), the Cuban green anole (*A. porcatus*), which has a pinkish purple dewlap, and the Hispaniolan green anole (*A. chlorocyanus*), which has a black and blue dewlap. The short spiny vertebral crest present in *A. garmani* is absent in these other anoles.

**History of introduction and current distribution:** *Anolis garmani* is native to Jamaica. Its presence in South Miami dates back to at least 1975 (Wilson and Porras, 1983), but the origin of this colony is unknown. This species has persisted also in a small area of Ft. Myers, Lee County, since 1985. The Ft. Myers colony resulted from an intentional introduction by people whose houses border a golf course. The Florida Fish and Wildlife Conservation Commission (1999–2002) reports that this species was established for a time in Martin County. The population, which was established in 1986 on a reptile dealer's property, died out in 1991 due to freezing temperatures.

**Habitat and habits:** This giant anole is typically classified as a canopy anole. In Florida, this species is found in urban South Miami on the trunks and limbs of large

trees at heights ranging from 0.5 to 12 m
above the ground. The anole perches conspicuously, often adopting a head-down position.
This species is active throughout the year, especially during the early afternoon hours.

**Reproduction:** Like other anoles, this species
is an egg layer. Percent left testis length
(6.4%) and width (4.2%) was recorded for a
male (95-mm SVL) captured in July 1997.

**Diet:** *Anolis garmani* is an omnivore. We
have reports of this species eating the brown
anole (*A. sagrei*) and fruit.

**Predators:** We know of no reports of predation in Florida.

*Anolis garmani*

## *Anolis porcatus* Gray, 1840

**Common name:** Cuban Green Anole

**Other common name:** Chameleo Verde

**Description:** Body color ranges from bright green with creamy vermiculations to
dark brown in metachrosis. The dewlap color is pinkish purple-mauve.

**Body size:** The largest male (76-mm SVL) and female (65-mm SVL) were from
North Miami, Dade County (Meshaka et al., 1997a).

**Similar species:** *Anolis porcatus* is easily confused with its close relative, the native green anole (*A. carolinensis*). The skull of *A. porcatus* is rugose with two
prominent frontal ridges that run lengthwise down the snout. Individual *A. carolinensis* have varying degrees of frontal ridges, usually much smaller than those of
*A. porcatus,* or the ridges are absent altogether. Definitive identification requires

*Anolis porcatus* Cuban Green Anole eating *A. distichus* Bark Anole. Photo by T. Lodge.

that an observer count the lamellae on the underside of the toes. *Anolis porcatus* tends to have more lamellae on the third and fourth toes on the front foot and the third toe on the hind foot than does *A. carolinensis* (Collette, 1961). The number of lamellae seems to reflect a function of body size and sex. On the Florida mainland, females with 20 or fewer lamellae on the third hind toe are *A. carolinensis,* whereas those with 21 or more lamellae on this toe are *A. porcatus.* Males with 22 or fewer lamellae on the third hind toe are *A. carolinensis,* and those with 25 or more lamellae on this toe are *A. porcatus.* These numbers are based on a limited number of specimens, and all of the *A. porcatus* were collected from Cuba. These two species are very closely related, presenting the possibility that they may be interbreeding, resulting in hybrids with lamellae number intermediate to those presented here. We refer those interested to the paper by Collette (1961).

An exotic species similar to *A. porcatus* is the Hispaniolan green anole (*A. chlorocyanus*) which has a blue and black dewlap. The Jamaica giant anole (*A. garmani*), also green in body color, has a vertebral crest and a brownish yellow dewlap.

**History of introduction and current distribution:** *Anolis porcatus,* endemic to Cuba, was discovered in a collection of *A. carolinensis* from Key West, Monroe County (Allen and Slatten, 1945). It has occurred at a site in North Miami, since 1991 and at sites in South Miami, Dade County, adjacent to that of *A. garmani*

since 1987 (Meshaka et al., 1997a). The de-
rivation of both colonies is unknown. *Anolis
porcatus* also occurs in Dade County in Coconut Grove and the King's Creek division of
Kendall. The complete geographic range of
*A. porcatus* in southern Florida has yet to be
defined. This anole may go undetected in
some areas because of its similarity in general appearance to *A. carolinensis*.

**Habitat and habits:** *Anolis porcatus* occupies the midtrunk and canopy of trees. In
Florida, this anole often occupies trees,
bushes, and the exterior of buildings in disturbed settings. Most individuals were observed above 2 m from the ground (Meshaka,
1999a, 1999b; Meshaka et al., 1997a). Similar to populations in Cuba, the North Miami
colony shares the habitat with other Cuban
species: the knight anole (*A. equestris*),
brown anole (*A. sagrei*), greenhouse frog
(*Eleutherodactylus planirostris*), and Cuban
treefrog (*Osteopilus septentrionalis*) (Meshaka et al., 1997a).

*Anolis porcatus*

**Reproduction:** On average, males (66.2-
mm SVL) are larger than females (55.6-mm
SVL). In July, a female (45.3-mm SVL) was
collected with a shelled egg (10.3 × 5.2-mm) and an enlarged follicle (7.3 ×
5.1-mm). Another female was collected with a shelled egg (10.8 × 5.5-mm) and
enlarged follicle (5.8 × 4.2-mm). The smallest juvenile (25-mm SVL) was also
collected in July (Meshaka et al., 1997a).

**Diet:** *Anolis porcatus* in Florida has a broad diet, feeding especially on flies
(Diptera) and ants (Hymenoptera). Its diet also includes *A. sagrei* and the bark
anole (*A. distichus*), as seen in the accompanying photo. Many of its prey are associated with trunk and canopy areas of trees (Meshaka et al., 1997a). A small
sample from south Miami ate primarily ants (Hymenoptera), beetles (Coleoptera),
and spiders (Aranea).

**Predators:** Although we have not yet documented predation on *A. porcatus* in
Florida, *A. equestris,* a predator of this species in Cuba (Meshaka et al., 1997a), is
syntopic with *A. porcatus* at all known sites.

## *Anolis sagrei* Duméril and Bibron, 1837

**Common name:** Brown Anole

**Other common name:** Cuban Anole

**Description:** Body color ranges from light brown to dark brown in metachrosis. The color of the dewlap is red or reddish orange with a light (cream or yellow) border. Many males have a nuchal and dorsal crest. Occasionally, males are encountered with a crest on the tail. Females and juveniles have a scalloped pattern on the dorsum.

**Body size:** The largest male (65-mm SVL) and female (48-mm SVL) were from the Dry Tortugas, Monroe County.

**Similar species:** This species most closely resembles the Puerto Rican crested anole (*Anolis cristatellus*), which has a reddish orange dewlap with a greenish yel-

*Anolis sagrei* Brown Anole. Photo by S.L. Collins.

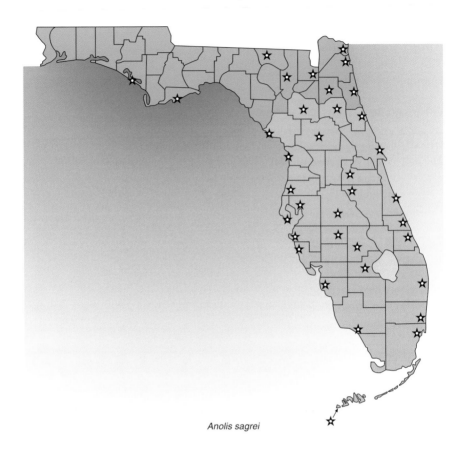

*Anolis sagrei*

low center and a fully developed dorsal crest extending through the tail. The dewlap of the largehead anole (*A. cybotes*) is creamy yellow in color.

**History of introduction and current distribution:** *Anolis sagrei* of the West Indies was first reported in Florida (Garman, 1887) from the Florida Keys, Monroe County. It was later reported from Hillsborough (Oliver, 1950b), Palm Beach (Oliver, 1950b; King, 1960), and Pinellas (Oliver, 1950b) counties. Other early county records include Dade (Bell, 1953), Broward (King 1960), and Lee (Ruibal, 1964). *Anolis sagrei* has been very successful in Florida and now occurs throughout most of peninsular Florida (Conant and Collins, 1998). Recently, this lizard has been reported from Bay (Means, 1990a) and Franklin (Means, 1996a) counties in the panhandle and in Nassau (Campbell and Hammontree, 1995), Baker, Columbia, Flagler, Hamilton, Putnam, Volusia (Campbell, 1996), and Clay (Townsend and Lindsay, 2001) counties in the northern peninsula. Other county records are from Collier (Conant, 1975), Glades (Corwin et al., 1977), St. Johns

(Meylan, 1977a), Indian River, St. Lucie (Myers, 1978c), Lee, Levy (Funk and Moll, 1979), Alachua (Wygoda and Bain, 1980), Highlands, Manatee, Marion, Osceola, Orange, Pasco, Polk, Sarasota (Godley et al., 1981), Duval (Lee, 1985), Brevard (Cochran, 1990), Citrus (Stevenson and Crowe, 1992b), and Hardee (Christman et al., 2000).

This anole was suspected to have initially arrived in Florida incidentally with cargo (Wilson and Porras, 1983), and nearly all of the early colonies were associated with seaports (King, 1960). The subsequent, rapid colonization throughout the state is attributed to dispersal through human agency (Godley et al., 1981; Campbell, 1996). Godley, et al. (1981) hypothesized vehicular rafting and transport in ornamental vegetation as the major modes of the anole's dispersal in Florida. Campbell (1996) found it at the northbound accesses of interstate rest areas and welcome centers but found none at southbound accesses. This observation likewise supports a hypothesis of vehicular rafting, but with a directional element. Attesting to its high vagility, we provide three extreme examples of dispersal by *A. sagrei* in ornamental plants: the presence of individuals in a nursery in Ohio; the discovery by a florist in Henderson, Tennessee, of hatchling *A. sagrei* in a floral arrangement that had arrived from Florida; and two specimens that arrived in a shipment of ornamental plants (from Florida) to a K-Mart in Brookings, South Dakota.

A series of studies on the phenetics and morphometrics of *A. sagrei* (Lee, 1985, 1987, 1992) confirm that no founder effect occurred in the colonization of *A. sagrei* in Florida, such that morphological variation in Florida is less than, and meristic variation does not exceed that of, West Indian and Central American populations. These studies confirm Oliver's (1950b) contention that a Tampa population was of Cuban origin. At a broader level, Lee (1985, 1987, 1992) confirms a predominantly Cuban influence in *A. sagrei*, far more so than Bahamian influence, in its early colonization of Florida. Interestingly, despite strong Cuban affinities, *A. sagrei* of Florida is now a distinct morphological entity of Florida. Selective pressures to account for this remain an interesting question. Genetic information indicates recent but separate colonizations from Havana by Miami and Tampa populations of *A. sagrei* (Lieb et al., 1983).

**Habitats and habits:** *Anolis sagrei* is a trunk-ground lizard in its habitat use (Meshaka, 1999a, 1999b; Table 10), often observed perched in open sunny areas. This anole perches on a wide variety of natural and human-made structures including trees, bushes, fences, and buildings. Its reported colonization of tropical hardwood hammocks seems to be marginal in scope; found along trails with high rates of human visitation (Dalrymple, 1988). Males are more often found on higher perch sites than females and juveniles (Meshaka, 1999a, 1999b; Table 10). Frequency of basking is similar between the sexes (Doan, 1996).

*Anolis sagrei* is well adapted to habitats modified by humans. For example, individuals have been observed copulating in the middle of a parking lot devoid of vegetation (Wilson and Porras, 1983). Additionally, we have observed individuals foraging at night under garden lamps, even in the rain. Their high tolerance to humans and human modified habitats has undoubtedly facilitated their successful colonization of Florida.

A comparison of basking behavior of *A. sagrei* and the bark anole (*A. distichus*) (Doan, 1996) revealed strong association in basking frequency between species and particularly between males of both species. Doan (1996) hypothesized that the similarity in food guild of the two species predisposed them to compete on other niche axes. Competition was predicted to be stronger for males than females because of the males' greater need to bask and therefore partition less time to feeding.

Local population densities can be high around buildings and other disturbed areas. Densities of over 0.97 lizards/$m^2$ are not uncommon in southern Florida (Lee et al., 1989). These densities are comparable to densities reported from the Bahamas (Schoener and Schoener, 1980). In the Everglades, it is most abundant in disturbed settings (Dalrymple, 1988).

**Reproduction:** In a sample from the Dry Tortugas, Monroe County, males (57.5 ± 4.5-mm SVL; range = 48–65; N = 19) were larger than females (45.1 ± 1.8-mm SVL; range = 42–48; N = 7). In Miami, Dade County, males were reproductive at 39-mm SVL, and females were reproductive at 34-mm SVL (Lee et al., 1989). In Miami, Lee et al. (1989) found reproduction to be strongly seasonal, with an activity peak from April through June. Periods with the fewest reproductively active individuals were from November through January for males and November through February for females. In Dade County, copulating pairs have been observed in August and September. In Homestead, Dade County, hatchlings were seen on the ground near the bases of trees as late as October and mature within 1 year of life.

**Diet:** This anole is primarily an insectivore, but it also eats hatchling green anoles (*A. carolinensis*) (Campbell and Gerber, 1996) and possibly *A. distichus*

**Table 10.** Perch heights of male *Anolis sagrei* from Red Road in Dade County.

| Perch Height (cm) | South Miami (Red Road) |
|---|---|
| 0–60 | 5 |
| 61–45 | 7 |
| 105–180 | 2 |
| 180+ | 0 |
| | Total: 14 |

(T. Lodge, pers. comm., 1996). Predation on hatchlings of *A. carolinensis* by *A. sagrei* is severe enough to result in the rapid decline of syntopic populations of *A. carolinensis,* the negative impacts of which are only somewhat amelio-rated by the presence of dense patches of understory (Campbell, 2000; Gerber and Echternacht, 2000). *Anolis sagrei* is cannibalistic (Nicholson et al., 2000).

**Predators:** Predators include the Cuban treefrog (*Osteopilus septentrionalis*) (Meshaka, 2001), knight anole (*A. equestris*), *A. carolinensis,* (Campbell, 2000), Jamaican giant anole (*A. garmani*), Cuban green anole (*A. porcatus*), northern curlytail lizard (*Leiocephalus carinatus armouri*) (Callahan 1982), American crow (*Corvus brachyrhynchos*) (K. Butterfield and E. McDuffee, pers. comm., 1999), corn snake (*Elaphe guttata*) (Wilson and Porras, 1983), broad-winged hawk (*Buteo platypterus*), and cattle egret (*Bubulcus ibis*). We have found *A. sagrei* eggs in the stomachs of a ringneck snake (*Diadophis punctatus*) and giant ameiva (*Ameiva ameiva*).

# SUBFAMILY CORYTOPHANINAE

## *Basiliscus vittatus* Wiegmann, 1828

**Common name:** Brown Basilisk

**Other common name:** Striped Basilisk

**Description:** The body color of males is olive brown with yellow dorsolateral stripes. That of females and juveniles is brown with yellow or cream dorsolateral stripes and dark brown to black crossbands. The body, tail, and legs are attenuated. Males have a very prominent crest on the back of the head and a low dorsal crest along the back and onto the tail. Females have a much smaller crest on the back of the head and lower dorsal crest on the body.

**Body size:** The largest male (160-mm SVL) and female (110-mm SVL) were from Kendall, Dade County.

**Similar species:** This long-limbed species is unlikely to be confused with any other in Florida.

**History of introduction and current distribution:** *Basiliscus vittatus* is native to Mexico and Central America southward to Colombia. This lizard was known from Dade County since 1976 and was also reported from Davie, Broward County (Wilson and Porras, 1983). Currently, populations of *B. vittatus* exist in Loxahatchee,

*Basiliscus vittatus* Brown Basilisk. Photo by W.E. Meshaka, Jr.

Palm Beach County, in Davie and Ft. Lauderdale, Broward County, and at Kendall, Tamiami, North Miami, South Miami, and near the Miami International Airport, Dade County. Colonies of *B. vittatus* were derived from releases and escapes of individuals from pet dealers and pet owners (Wilson and Porras, 1983). Subsequently, *B. vittatus* has dispersed throughout much of the greater Miami area along canals.

**Habitats and habits:** *Basiliscus vittatus* is generally found near dense vegetation and water of borrow pits and canals. Adults perch in trees or bask on the ground; however, juveniles are frequently seen along the shoreline near dense vegetation. Members of all size-classes can be found within 75 m of water on the ground in open areas near bushes and trees. At night, individuals sleep on tree branches.

*Basiliscus vittatus* is most active on hot sunny days but may be seen basking on cool or overcast days. During relatively cooler periods, this lizard often basks in open areas such as sidewalks.

This lizard and others in the genus are well known for their ability to run on their hind legs, even across the surface of water. In Florida, startled individuals retreat up trees, flee on the ground into dense vegetation, dive under the water, or run on the surface of water. Juveniles enter the water more readily than adults.

**Reproduction:** Gravid females have been found in March (D. Holley, pers. comm., 1993). We have also collected gravid females and observed hatchlings in

June and July. One female, 98.8-mm SVL, collected in mid-July, contained four oviductal eggs (20.3 ± 0.7-mm in length and 10.6 ± 0.4-mm in width). Gravid females, collected in June, laid eggs in July (P. Bedsole, pers. comm., 1995). One of these females produced a second clutch of six eggs on 10 October 1996. These six eggs were incubated at a temperature of 28°C and four hatched on 11 December 1996. These hatchlings measured 35–36-mm SVL. The left testis of a large male (160-mm SVL) collected in May measured 9.5 x 6.4-mm.

**Diet:** *Basiliscus vittatus* is primarily carnivorous. Beetles (Coleoptera), roaches (Dictyoptera), ants (Hymenoptera), true bugs (Hemiptera), and ficus fruits were recovered from the stomachs of a small sample of *B. vittatus* collected along a canal in Dade County (Table 11). Recent captives will readily accept domestic crickets, mealworms, and newborn mice.

**Predators:** The corn snake (*Elaphe guttata*), eastern racer (*Coluber constrictor*), and eastern indigo snake (*Drymarchon corais couperi*) prey on young *B. vittatus* (S. St. Pierre, pers. comm., 1995). *Elaphe guttata* have been ob-

*Basiliscus vittatus*

**Table 11.** Diet of three male and three female *Basiliscus vittatus* from Dade County. Numbers indicate number of prey (number of lizards consuming each prey category).

| Taxa of Food Items | Male | Female | Total |
|---|---|---|---|
| Coleoptera | 8(3) | 6(3) | 14(6) |
| Dictyoptera | 2(2) | 1(1) | 3(3) |
| Hymenoptera (Formicidae) | 0(0) | 4(1) | 4(1) |
| Hemiptera | 1(1) | 0(0) | 1(1) |
| Ficus Fruit | 1(1) | 6(1) | 7(2) |

served eating juvenile *B. vittatus,* one after another, that had been sleeping next to each other on branches suspended over canals (S. St. Pierre, pers. comm., 1995).

---

# FAMILY GEKKONIDAE

## *Cosymbotus platyurus* (Schneider, 1792)

**Common name:** Asian House Gecko

**Other common name:** Flat-Tailed Gecko

**Description:** The dorsal color is brownish gray with a faint pattern. Webbing is present between the toes, and a fold of skin lies along each side of its body.

**Body size:** The largest male (61.2-mm SVL) was from Homestead, Dade County. The largest female (48.6-mm SVL) was from Clearwater, Pinellas County.

**Similar species:** The fold of skin along the side of the body distinguishes this species from the superficially similar hemidactyline geckos established in Florida.

*Cosymbotus platyurus* Asian House Gecko. Photo by R.D. Bartlett.

**History of introduction and current distribution:** *Cosymbotus platyurus* is native to southern China, Sri Lanka, northern India, Thailand, Malaysia, the Indoaustralian Archipelago to New Guinea, and the Philippines (Zhao and Adler, 1993). The first known colony of this gecko in Florida is from Clearwater, and the colony had not expanded from the original site during the 10 years of its existence since the mid-1980s (Meshaka and Lewis, 1994). We discovered other colonies of *C. platyurus* in Gainesville, Alachua County and Ft. Myers, Lee County, in 1995; and Homestead in 1996. The Gainesville population, located at the former site of a reptile importer, was still in existence in 1999 (Hauge and Butterfield, 2000a).

All of these colonies exist on or near buildings previously occupied by pet dealers. Therefore, the likely founders of all of these colonies were individuals that had escaped from or were released by the pet dealers. The Ft. Myers colony may have died out, as no individuals were observed during two visits to the site in 1997. We also have reports of a population in Tampa, Hillsborough County.

*Cosymbotus platyurus*

**Habitat and habits:** In Florida, *C. platyurus* has been found on buildings and associated human structures but has not been observed on vegetation. The relative abundance of this gecko at Clearwater was measured over a period of 15 years using visual surveys. These surveys yielded 40–50 individuals per 0.5 hour (Meshaka and Lewis, 1994).

This species was abundant at the Ft. Myers site in 1995, where it co-occurred with the house gecko (*H. frenatus*), Indo-Pacific gecko (*H. garnotii*), and ringed wall gecko (*Tarentola annularis*). In 1997, no individuals of *C. platyurus* were observed, but *H. frenatus* was more abundant than in 1995. Therefore, the presence of other exotic species may negatively impact populations of *C. platyurus*.

**Reproduction:** Gravid females have been found in April, and Meshaka and Lewis (1994) found hatchlings during a census in November. The left testis of a male (61.2-mm SVL) from Homestead collected in July 1996 measured 5.1 × 3.0 mm.

**Diet:** This species is an insectivore. One fly (Diptera) was recovered from the stomach of an adult collected in Homestead. We have no other diet information for Florida specimens.

**Predators:** *Tarentola annularis* preys on *C. platyurus*. The Cuban treefrog (*Osteopilus septentrionalis*) and tokay gecko (*Gekko gecko*), two predators of lizards that also share the same habitat, are potential predators of this gecko.

---

## *Gekko gecko* (Linnaeus, 1758)

**Common name:** Tokay Gecko

**Other common name:** none

**Description:** The body coloration is blue-mauve with small red and orange spots throughout. The skin texture is warty.

**Body size:** This is by far the largest of the introduced geckos. The largest male (172-mm SVL) and female (142-mm SVL) were from Homestead, Dade County.

**Similar species:** The large size and unique coloration of this gecko make it unlikely to be mistaken for any other species in Florida.

**History of introduction and current distribution:** *Gekko gecko* of southeastern Asia was first known from Dade and Broward counties (Wilson and Porras, 1983). Small disjunct colonies occurred on buildings in scattered locations. Its initial introduction, perhaps during the 1960s, and its disjunct pattern of dispersal are linked with animal dealers and pet owners. The primary dispersal agents, pet shops, and secondary dispersal agents, consumers, can cause rapid but often isolated and discontinuous geographic expansion. The colonization around human dwellings can occur rapidly, as evidenced by a colony in Tallahassee, Leon County (Means, 1996b). We also report a second colony at a pet shop and surrounding trees in Leon County. A single juvenile specimen was recorded from the former site of a reptile importer in Gainesville, Alachua County (Butterfield and Hauge, 2000). We have also collected this species from buildings and trees on Key West, Monroe County. In addition to these records, Bartlett and Bartlett (1999) report the species from Lee, Hillsborough, Palm Beach, Broward, and Collier counties. We also have reports, with voucher specimens or photographs, of populations from Pinellas and Highlands counties.

**Habitat and habits:** *Gekko gecko* occupies buildings, piles of concrete blocks, and other human structures. This gecko has also been observed in the crevices of

*Gekko gecko* Tokay Gecko (adult). Photo by R.D. Bartlett.

*Gekko gecko* Tokay Gecko (eggs and hatchlings). Photo by S.D. Marshall.

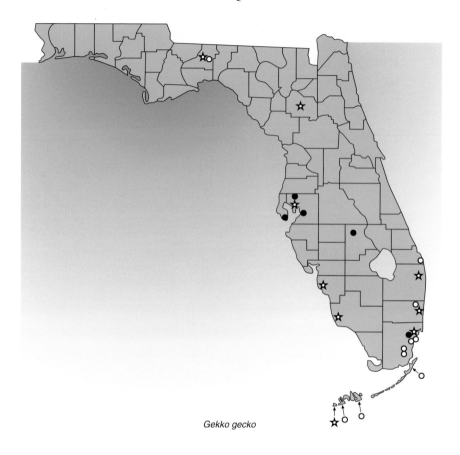

*Gekko gecko*

ficus trees in urban and heavily disturbed areas, including hammocks in Dade and Monroe counties.

*Gekko gecko* is active at night when the ambient temperature is above 15°C. Individuals have been observed on the exterior walls of buildings, ficus trees, black olive trees, wooden telephone poles, and in one case, a parked van. During the day we have seen this species peeking out of the crevices of ficus trees. On sultry days and just before hard rains, it may emerge from its diurnal retreat. We also have observed individuals basking on ficus trees during the middle of the day in mid-January.

Males of this species emit a loud, two-syllable call, described as sounding like the word "to-kay." A series of these two-syllable calls are usually uttered, the final call often lengthened and fading. At Dade County sites in Homestead and South Miami, males have been heard calling at night during the winter and spring. Captive individuals from the Miami International Airport site called most intensely just before sunrise.

**Table 12.** Nest parameters of *Gekko gecko* from Dade County. * = from the same nest site.

| Date | Location | Height (cm) | Number of Eggs Hatched/Unhatched |
|------|----------|-------------|----------------------------------|
| 5 May 1992 | Ficus | 250 | 5/7 |
| | Ficus | 275 | 6/33 |
| | Ficus* | 342 | 7/13 |
| | Ficus | 350 | 1/3 |
| | Building | 180 | 20/120 |
| 8 July 1992 | Ficus | 208 | 0/5 |
| | Ficus | 231 | 0/4 |
| | Ficus* | 342 | 0/4 |
| 18 January 1993 | Ficus* | 342 | 5/0 |

Local populations can be large. For example, about 20 individuals were counted on the walls of a pet shop in Tallahassee, Leon County, during a period of approximately 1 hour, at night during May. Additionally, 20 individuals were captured at night in approximately 2.17 man-hours of search near the Miami International Airport (Meshaka et al., 1997b). Wilson and Porras (1983) suggest that there exists a strong sense of philopatry by this territorial lizard.

**Reproduction:** Both sexes mature at about 110-mm SVL; however, on average, adult males (144.9 ± 10.9-mm SVL; range = 126–166; N = 14) are larger than females (126.8 ± 7.3-mm SVL; range = 110–138; N = 18), as measured from more than one site in southern Florida (Meshaka et al., 1997b). Collections between May and September have yielded males, gravid females, and a juvenile (50-mm SVL). One female (131-mm SVL) contained two round (20 mm) shelled eggs. A smaller female (121-mm SVL) also contained two round (16, 17 mm) shelled eggs. Eggs were found 1.8–3.4 m above the ground in communal nests located in the cavities of ficus trees and on buildings (Table 12). We have observed communal nests that contained up to 140 eggs. One exceptionally large nest was located above the ground behind an electrical meter box and was attended by a male. After collecting the male and later surveying the site for a second time, we observed a different male attending the eggs. Mean percent length and width of the left testis

**Table 13.** Diet of 18 *Gekko gecko* from two sites in Dade County. Numbers indicate number of prey (number of lizards consuming each prey category).

| Taxa of food items | Total |
|---|---|
| Arachnida (Aranea) | 2(1) |
| Coleoptera | 14(10) |
| Dictyoptera | 16(12) |
| Dermaptera | 1(1) |
| Diptera | 1(1) |
| Hemiptera | 4(3) |
| Hymenoptera (Formicidae) | 1(1) |
| Lepidoptera | 10(6) |
| Orthoptera | 6(3) |
| Isopoda | 5(1) |
| *Hemidactylus mabouia* | 1(1) |

from eight males collected in July was $4.6 \pm 0.8\%$ (range = 3.3–5.9) × $2.5 \pm 0.3\%$ (range = 2.0–2.9).

**Diet:** The diet of a combined sample of *G. gecko* from sites in Homestead and near the Miami International Airport, Dade County, consisted primarily of roaches (Dictyoptera), beetles (Coleoptera), moths (Leptidoptera), and moth caterpillars (Table 13). This gecko also preyed on the tropical gecko (*Hemidactylus mabouia*), which occurred at both sites. Meshaka and colleagues (1997b) examined a sample of *G. gecko* stomachs collected only from trees near the Miami International Airport. Stomachs from these geckos contained prey that was common on trees and surrounding leaf litter such as roaches and lepidopteran larvae. Love (2000) documented a large *G. gecko* consuming a juvenile corn snake (*Elaphe guttata*). *Gekko gecko* probably preys on other exotic geckos.

**Predators:** The Cuban treefrog (*Osteopilus septentrionalis*), a predator of geckos and other vertebrates (Meshaka, 2001), is a potential predator of hatchlings.

---

## *Gonatodes albogularis fuscus* (Duméril and Bibron, 1836)

**Common name:** Yellowhead Gecko

*Gonatodes albogularis fuscus* Yellowhead Gecko. Photo by R.D. Bartlett.

**Other common name:** none

**Description:** The body color of males is dark grayish blue. Males have a bright yellow head. Females and juveniles are speckled brown.

**Body size:** The largest male (40.3-mm SVL) and female (37.8-mm SVL) were both from Key West, Monroe County (Duellman and Schwartz, 1958). Unsexed adults examined from museum collections ranged 31.3–40.2-mm SVL.

**Similar species:** The unique body color (especially of males) and diurnal habits distinguish *G. a. fuscus* from the other geckos established in Florida.

**History of introduction and current distribution:** *Gonatodes albogularis fuscus* is native to Central America, South America, and the West Indies. This gecko was first detected in Florida from Key West more than 60 years ago (Carr, 1939). Here, it was very numerous but apparently restricted to railroad debris and adjacent buildings on the western end of the is-

*Gonatodes albogularis*

land. Initial introduction appears to have been accidental and associated with commerce in the Caribbean. The gecko was still thriving at the time of a study by Duellman and Schwartz (1958), although less commonly in the Navy yards and downtown Key West. Wilson and Porras (1983) reported that a large series was secured from Key West in 1971, but the population has since declined. Lawson et al. (1991) collected two specimens and observed others in a vacant lot on Key West, and we know of a small colony on a large wild tamarind tree on private land on Key West. Another isolated but thriving colony existed on Stock Island until many of the large ficus trees were removed during landscaping in the mid-1990s. We have also seen this species on a ficus tree near the Truman Annex on Key West. From these records, *G. a. fuscus* apparently remains established on the very lowermost Florida Keys, although clearly it is not as abundant as it was in the past.

On mainland Florida, *G. a. fuscus* may no longer occur in Coconut Grove, Dade County (Wilson and Porras, 1983), where it was first reported (King and Krakauer, 1966). The Haitian subspecies, *G. a. notatus,* has been reported from a site in Parkland, Broward County, where one or two individuals have been observed since 1993 on a ficus tree (Bartlett and Bartlett, 1995). The site was at one time a pet dealership but is now a residence and horse stable.

**Habitat and habits:** Carr (1940) described this species as a building dweller on Key West, but it also occurs on trees. Unlike most geckos, *G. albogularis* is active during the day. At midday, individuals have been found upside-down on horizontal tree branches (Bartlett and Bartlett, 1995). Additionally, we record a male having foraged for 20 minutes on the trunk and aerial root systems of a ficus tree at dusk. During this time, the gecko wagged the tip of its tail continuously.

**Reproduction:** This species is an egg layer; however, we have no information on reproduction for Florida populations.

**Diet:** We have not surveyed food habits for Florida populations. The species is known to eat a variety of arthropods.

**Predators:** We know of no reports of predation on this species in Florida populations.

---

## *Hemidactylus frenatus* Duméril and Bibron, 1836

**Common name:** House Gecko

**Other common names:** Chit Chat, Tropical House Gecko

*Hemidactylus frenatus* House Gecko. Photo by R.D. Bartlett.

**Description:** The dorsal color is gray with faint longitudinal dark stripes, and the venter is white. The body texture is smooth. Femoral pores are present.

**Body size:** The largest male (53.0-mm SVL) was from Key West, Monroe County, and the largest female (52.5-mm SVL) was from Homestead, Dade County.

**Similar species:** The Indo-Pacific gecko (*Hemidactylus garnotii*), which this species most closely resembles, has a distinctly yellow venter and a yellowish orange ventral aspect of the tail. The tropical gecko (*H. mabouia*) has a moderately warty body and chevron-shaped markings. The Mediterranean gecko (*H. turcicus*), superficially similar to *H. frenatus,* lacks femoral pores and has many tubercles on its skin. The Asian house gecko (*Cosymbotus platyurus*) has a fold of skin along each side of its body.

**History of introduction and current distribution:** *Hemidactylus frenatus,* from southeast Asia, was detected in Monroe County in 1993 (Meshaka et al., 1994b). This gecko was reported as abundant on buildings on Key West and Stock Island. Its introduction appears to have been related to the trade of exotic pets involving

a particular pet shop on Key West. Populations subsequently detected in Ft. Myers,
Lee County, and Homestead are also associated with individuals released from pet
shops.

**Habitat and habits:** *Hemidactylus frenatus*
occupies buildings and we have observed individuals on ficus trees. This gecko is active
at night when temperatures are warm. One of
its common names, the chit chat, is derived
from the sound of its call.

 *Hemidactylus frenatus* occurs with *H.
mabouia* on Key West and Stock Island. In
1993 both species occurred in equal numbers
on Key West, but *H. frenatus* was outnumbered 2:1 by *H. mabouia* on Stock Island
(Meshaka et al., 1994b). By 1995, *H.
mabouia* greatly outnumbered *H. frenatus*
on Stock Island. During surveys in 1997, *H.
frenatus* was still present on both islands but
was outnumbered by *H. mabouia*. Consequently, Meshaka and coworkers (1994b)
appear to have been correct that only one
hemidactyline at a time can stably exist on
buildings in Florida but it appears that it is *H.
mabouia* that is moving southward and not
*H. frenatus* that is moving northward. On the

*Hemidactylus frenatus*

other hand, this species has replaced *Cosymbotus platyurus* and greatly outnum-
bers *H. garnotii* within its range in Ft. Myers.

**Reproduction:** The body sizes of two males (50, 53-mm SVL) were similar to the
body sizes of five females (49.9 ± 2.4-mm SVL; range = 46.0–52.5). Females
collected in September were gravid. They may be capable of producing at least
four clutches (Meshaka et al., 1994b). A total of six shelled eggs from four females
measured 9.2 ± 0.4 mm; range = 8.6–9.6 × 7.3 ± 0.4 mm; range = 7.0–8.0. A
hatchling (23-mm SVL) was collected on Key West in September 1993.

**Diet:** A sample of stomachs of *H. frenatus* from Stock Island contained invertebrates, especially flies (Diptera) and roaches (Dictyoptera) (Table 14).

**Predators:** In Homestead, the ringed wall gecko (*Tarentola annularis*) preys on
this gecko. Potential predators include the Cuban treefrog (*Osteopilus septentrionalis*) and the tokay gecko (*Gekko gecko*), which share the same habitat.

**Table 14.** Diet of 12 *Hemidactylus frenatus* from Monroe County. Numbers indicate number of prey (number of lizards consuming each prey category).

| Taxa of Food Items | Total |
|---|---|
| Coleoptera | 1(1) |
| Dictyoptera | 4(4) |
| Diptera | 21(7) |
| Homoptera | 2(2) |
| Hymenoptera (other) | 3(2) |
| Lepidoptera | 2(1) |
| Unidentified Insect | 1(1) |

## *Hemidactylus garnotii* Duméril and Bibron, 1836

**Common name:** Indo-Pacific Gecko

**Other common name:** none

**Description:** The dorsal color ranges from yellowish brown to black with white flecks. The venter is yellow. The tail is dorsoventrally flattened, saw-toothed, and

*Hemidactylus garnotii* Indo-Pacific Gecko. Photo by S.L. Collins.

*Hemidactylus garnotii*

yellowish orange beneath. Adults often have calcium-filled chalky-white endo-lymphatic sacs visible on either side of the neck.

**Body size:** The largest adult (63.1-mm SVL) was from Palmdale, Glades County.

**Similar species:** The house gecko (*Hemidactylus frenatus*) closely resembles *H. garnotii* but has a white venter and ventral aspect to the tail. The tropical gecko (*H. mabouia*) has a slightly warty skin and chevron-shaped markings. The Mediterranean gecko (*H. turcicus*) has tubercles on its skin. The Asian house gecko (*Cosymbotus platyurus*) has a fold of skin along each side of its body.

**History of introduction and current distribution:** *Hemidactylus garnotii* is native to southeast Asia. Wilson and Porras (1983) knew of this gecko from two Hialeah, Dade County, locations from a few years prior to the first record in Miami and Coconut Grove, Dade County (King and Krakauer, 1966). The source of the Hialeah colonies is thought to be by incidental stowaways associated with commerce (Wilson and Porras, 1983). The Miami colony may have been the result of deliberate introductions associated with the International Indian Ocean Expedition during 1960–1963 (King and Krakauer, 1966). Kluge and Eckardt (1969) reported this species from Everglades National Park, where it still occurs (Meshaka, 2000; Meshaka et al., 2000). This species was not recorded in the Keys by Duellman and Schwartz (1958).

*Hemidactylus garnotii* is now widely distributed and many of the recently detected colonies were likely established long before their official record in this volume or elsewhere. Early county records were Sanibel Island in Lee (McCoy, 1972), Martin, Indian River, Brevard, St. Lucie (Myers, 1979), Collier (Mitchell and Hadley, 1980), Dry Tortugas in Monroe (Steiner and McLamb, 1982), Upper Matacumbe and Grassy keys (Wilson and Porras, 1983), and Orange (Smith, 1983). Recent county records include Palm Beach, St. Johns (Conant and Collins, 1991), Citrus (Stevenson and Crowe, 1992c), Pinellas (Crawford and Somma, 1993b), Volusia (Reppas 1999), Hardee (Christman et al., 2000), and Flagler County (Lindsay and Townsend, 2001). Additional records based on our observations in 1993 are Carrabelle (Franklin County), Lake Placid (Highlands County), Manatee County (near I-75), Stuart and Hutchinson Island (Martin County), Ft. Drum and Okeechobee (Okeechobee County), Holopaw and Yee Haw Junction (Osceola County), and Sarasota (Sarasota County). We also note from museum records individuals from LaBelle (Hendry County) in 1987 and Tampa (Hillsborough County) since the 1960s. A record from Key West in Monroe County (Meshaka et al., 1994a) also exists.

The rapid and haphazard pattern of its dispersal is human mediated, including transport via the foliage trade (Meshaka, 1996a). Transport on building materials is another dispersal agent of this species. For example, during a 2-year study in Everglades National Park, three individuals (40, 54, 54-mm SVL) were observed during 3 months of construction on a boardwalk near Flamingo, Monroe County (Meshaka, 2001).

**Habitat and habits:** *Hemidactylus garnotii* is most often found on buildings (Wilson and Porras, 1983), where it is conspicuous and often abundant (Meshaka, 2000, 2001). However, this gecko is not restricted to buildings; individuals are routinely found in palm boots (Meshaka, 1996a), beneath pine bark in isolated pinelands of Homestead, Dade County, beneath Australian pine bark along Lake Okeechobee, and at night on ficus and assorted palm trees throughout southern Florida. In Everglades National Park, *H. garnotii* has been observed in hammocks and mangrove forest (Meshaka, 2000).

*Hemidactylus garnotii* is generally considered to be nocturnal, but we have frequently observed it foraging on overcast days and occasionally on clear days. Evening activity usually begins at dusk, when it emerges from its diurnal retreat to forage. Individuals perch within 1 m of green anoles (*Anolis carolinensis*) under the eaves of old buildings in Lake Placid. This species is active during a wide range of physical conditions. This species is active in Everglades National Park under the following conditions: mean temperature of 25.1°C (16–30°C), mean relative humidity of 84.7% RH (65–100), and mean rainfall of 1.1-cm rainfall (0.0–12.9) (Meshaka, 2000). Frankenburg (1984) noted that individuals emerge from retreats before dark (1700 hrs) and are active until 0200 hrs.

This gecko apparently does not defend a territory or form groups but is widely dispersed and will fight other geckos that approach too closely. *Hemidactylus garnotii* fights more frequently with conspecifics than it does with *H. turcicus*; however, it is more frequently the winner of those interspecific battles (Frankenberg, 1982a; 1984).

The geographic range expansion of this gecko in southern Florida during the 1970s and 1980s appears to have paralleled the decline of *H. turcicus* as far north as Tampa. On the other hand, once abundant in the Everglades, *H. garnotii* has within a few years been mostly replaced by *H. mabouia,* a recent and superior competitor on buildings (Meshaka, 2000). *Hemidactylus garnotii* has also been largely replaced on the Upper and Middle Keys and parts of southern mainland Florida by *H. mabouia*. On the lower Keys, colonization of *H. frenatus* and *H. mabouia* are associated with low population densities of *H. garnotii,* suggesting that *H. garnotii* arrived after *H. turcicus* but before *H. frenatus* and *H. mabouia*. *Hemidactylus frenatus* is also replacing *H. garnotii* on the west coast of Florida.

**Reproduction:** *Hemidactylus garnotii* is a triploid, all-female, parthenogenetic species (Kluge and Eckhardt, 1969), which reproduces continuously in southern Florida (Meshaka, 1994b). The smallest individuals are 22-mm SVL, and maturity is reached at the end of the first year at 48–49-mm SVL (Meshaka, 1994b). Shelled eggs range from 7.0 (Voss, 1975) to 10.2-mm (Meshaka, 1994b) and are laid singly or in pairs. At least three clutches are possible annually. Eggs of this species do not require absorption of water from the environment, and their water vapor conduction is low (Dunson, 1982), perhaps contributing to the colonization success of this species by increasing the range of physical conditions conducive to successful incubation.

**Diet:** A sample of *H. garnotii* collected from lighted buildings in Everglades National Park fed primarily on flies (Diptera, including mosquitoes) and various hymenopterans (Meshaka, 2000, 2001).

**Predators:** The Cuban treefrog (*Osteopilus septentrionalis*), knight anole (*Anolis equestris*), and corn snake (*Elaphe guttata*) prey on *H. garnotii*. Predation by *O. septentrionalis* has been shown to reduce population size and increase mean individual body size in populations of *H. garnotii*. For example, the mean SVL of adult geckos from buildings with *O. septentrionalis* was 57.8 ± 3.7-mm (range = 50.0–63.1; N = 30), whereas the mean SVL of geckos from buildings without *O. septentrionalis* was 55.3 ± 3.0-mm (range = 48.0–60.0; N = 37). Juveniles from those buildings with *O. septentrionalis* composed a smaller part of the collection (19%) than those from buildings without this predator (34%). This observation suggests that *O. septentrionalis* alters populations of *H. garnotii* primarily by consuming juveniles.

---

## *Hemidactylus mabouia* (Moreau de Jonnès, 1818)

**Common name:** Tropical Gecko

**Other common names:** African House Gecko, House Gecko, Wood Slave

**Description:** The body color ranges from brownish pink to dark brown with black dorsal chevrons. The tail has dark bands. This species can become very pale, almost white in appearance. The body is moderately warty. Femoral pores are present.

**Body size:** The largest males (67-mm SVL) were collected from Flamingo and Key West, Monroe County. The largest female (65-mm SVL) was collected near the Miami International Airport, Dade County.

**Similar species:** This species can be distinguished from the other hemidactyline geckoes in Florida by having black chevron-shaped markings. Its skin is more warty than that of the Indo-Pacific gecko (*H. garnotii*) or the house gecko (*H. frenatus*). The Mediterranean gecko (*H. turcicus*) lacks chevrons and femoral pores

*Hemidactylus mabouia* Tropical Gecko. Photo by R.D. Bartlett.

and tends to be more tuberculate. The Asian house gecko (*Cosymbotus platyurus*) has a fold of skin along each side of its body.

**History of introduction and current distribution:** *Hemidactylus mabouia* is native to Africa and has successfully invaded many tropical areas worldwide. In 1978, Smith and Koehler predicted this species to colonize North America. *Hemidactylus mabouia* was first recorded in Florida from Crawl Key, Monroe County (Lawson et al., 1991). Shortly thereafter it was found to be ubiquitous throughout the Florida Keys; Bahia Honda, Big Pine Key, Fat Deer Key, Key Vaca, Key West, Lower Sugarloaf Key, Middle Torch Key, Plantation Key, Stock Island, and Sunshine Key, (Günther et al., 1993; Meshaka et al., 1994a; Watkins-Colwell and Watkins-Colwell, 1995b). It was also found on Garden Key in the Dry Tortugas (Meshaka and Moody, 1996) and the southern mainland counties of Dade and Broward (Butterfield et al., 1993), where it dispersed northward into the counties of Glades (Meshaka et al., 1994c), Brevard (Criscione et al., 1998), and Orange (Butterfield et al., 2000). On mainland Florida, it is also present in Lake Harbor (Palm Beach County). In addition, the FLMNH lists specimens from Lee, Collier, Martin, Charlotte, and Okeechobee counties.

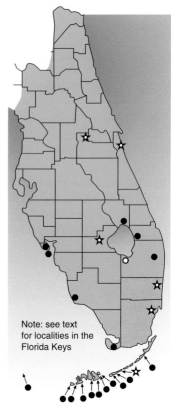

Note: see text for localities in the Florida Keys

*Hemidactylus mabouia*

Although the age of its introduction is unknown, its ubiquity (Meshaka et al., 1994a), superficial resemblance to *H. turcicus,* and ability to quickly displace other hemidactyline geckos (Meshaka, 2000; Meshaka and Moody, 1996) corroborates earlier suggestions by Meshaka and colleagues (1994a) that the introduction of *H. mabouia* in Florida could have occurred as recently as the early 1980s. Multiple introductions to multiple sites are likely, perhaps from agricultural shipments from Brazil and Puerto Rico, where *H. mabouia* is also an established exotic (Schwartz and Henderson, 1991). Lawson et al. (1991) proposed a connection for dispersal via pleasure craft between the Florida Keys and the Bahamas.

**Habitat and habits:** The highest population densities of this gecko are associated with buildings (Meshaka, 2000, 2001). However, *H. mabouia* is not re-

stricted to this habitat. Individuals have commonly been observed on black
olive, ficus, coconut palm, royal palm, and queen palm trees in urban areas and
disturbed lots. *Hemidactylus mabouia* is present in pinelands in Dade County
but has not been detected in natural systems of Everglades National Park (Meshaka, 2000).

*Hemidactylus mabouia* is primarily nocturnal, although during the day adults
occasionally bask on ficus trees a few centimeters from a retreat and sometimes
only a few centimeters from brown anoles (*Anolis sagrei*) and bark anoles (*A. distichus*). When active, *H. mabouia* seems to be more wary than *H. turcicus* but less
so than the *H. garnotii* and *H. frenatus*.

In disturbed areas of Everglades National Park, activity is highest in the wet
season (May-October). Individuals are active under a wide range of physical
conditions. In southern Florida, individuals are most active at a mean temperature of 25.1°C (16.0–30.0°C), mean humidity of 84.5% RH (60–100% RH), and
mean rainfall of 1.0 cm (0.0–12.9-cm) (Meshaka, 2000). *Hemidactylus
mabouia* is also more active in the wet season than in the dry season (Meshaka,
2000).

*Hemidactylus mabouia* quickly becomes locally abundant and displaces the
other hemidactylines. This species has replaced a population of *H. turcicus* on
Sugarloaf Key. In less than a 5-year period during the 1990s, the gecko community on a Sugarloaf Key hotel changed from dense populations of both *H. turcicus*
and the ashy gecko (*Sphaerodactylus elegans*) (R.D. Bartlett, pers. comm., 1997)
to a large population of *H. mabouia* and a small population of *S. elegans*. An early
study (Meshaka et al., 1994b) and a subsequent survey during 1997 at the same location revealed that the relative abundance of *H. mabouia* exceeded that of *H. frenatus*, suggesting some level of species replacement. *Hemidactylus mabouia* replaced *H. garnotii* at Garden Key, Dry Tortugas (Meshaka and Moody, 1996) and
is also replacing *H. garnotii* on the southeastern mainland of Florida. In only a 3-
year span, *H. mabouia* replaced *H. garnotii* on buildings in Everglades National
Park (Meshaka, 2000). During the 1990s, we have observed similar patterns of *H.
mabouia* replacing other hemidactylines in South Miami, Kendall, and most recently, in Homestead. The pattern is always the same, whereby *H. mabouia*
quickly saturates the site with more individuals than there had originally been of
*H. garnotii*, and *H. garnotii* simultaneously diminishes in population size. Occasionally, *H. garnotii* vanishes altogether.

**Reproduction:** Average adult body size (58–60-mm SVL) and body size at sexual maturity (49-mm SVL) are similar for both sexes. Both sexes are mature at
the end of their first year. Copulations in Homestead were observed at night in
December and March. No more than two eggs (7.0–10.7 mm) at a time are laid
throughout the year in southern Florida, and based on counts of ova >1.0 mm,
up to seven clutches can be produced each year (Meshaka et al., 1994a, 1994d;
Meshaka and Moody, 1996; Meshaka, 2000, 2001). In Miami we have found
multiple clutches of eggs deposited in the narrow crevices of ficus trees and in

the leaf axils of coconut palm trees. Communal clutches of up to 30 eggs were found under a discarded piece of carpet on Crawl Key (J. Lewis, pers. comm., 1993). We have found multiple clutches (hatched and unhatched) of *H. mabouia* with those of *Sphaerodactylus elegans* under rocks in a small and disturbed tropical hardwood hammock on Stock Island, in late March. We do not know the body size at the time of hatching, but the smallest individuals we have observed were 22.0-mm SVL.

**Diet:** The diet of *H. mabouia* is broad, including hard-bodied and soft-bodied prey of a variety of sizes (Meshaka and Moody, 1996; Meshaka, 2000, 2001). Stomach contents of a sample of individuals collected from Everglades National Park consisted primarily of flies (Diptera) and spiders (Aranea) (Meshaka, 2000, 2001). The stomach contents of a combined sample from a single site in Homestead and a single site near Miami International Airport, consisted primarily of moths (Leptidoptera), moth caterpillars, dermapterans (Dermaptera), and isopods (Isopoda) (Table 15).

**Predators:** *Hemidactylus mabouia* is preyed on by the Cuban treefrog (*Osteopilus septentrionalis*) and the tokay gecko (*Gekko gecko*).

**Table 15.** Diet of 23 *Hemidactylus mabouia* from Dade County. Numbers indicate number of prey (number of lizards consuming each prey category).

| Taxa of Food Items | Total |
|---|---|
| Arachnida (Aranea) | 3(2) |
| Coleoptera | 1(1) |
| Dictyoptera | 5(5) |
| Diptera | 1(1) |
| Hemiptera | 2(2) |
| Homoptera | 5(2) |
| Hymenoptera (Formicidae) | 1(1) |
| Lepidoptera | 9(8) |
| Orthoptera | 5(2) |
| Isopoda | 1(1) |
| Gastropoda | 2(2) |

## *Hemidactylus turcicus* (Linnaeus, 1758)

**Common name:** Mediterranean Gecko

**Other common name:** Turkish Gecko

**Description:** Body color ranges from pink to pale yellow with irregular brown spots. The body is covered with tubercles and appears very warty. Seven or eight precloacal pores are present. Femoral pores are absent.

**Body size:** The largest male (58-mm SVL) and female (57-mm SVL) were from the southeastern shore of Lake Istokpoga, Highlands County (Meshaka, 1995). The maximum body length of 79-mm SVL listed in Meshaka (1995) was a misprint.

**Similar species:** This is the only gecko in Florida with tuberculate skin. The tropical gecko (*Hemidactylus mabouia*) is similar in appearance to *H. turcicus*, but its skin is only slightly warty, it has dark dorsal chevrons (when it does not occasionally shift to pale white), and it has pronounced femoral pores. The house gecko (*H. frenatus*) and Indo-Pacific gecko (*H. garnotii*) have smooth skin. The Asian house gecko (*Cosymbotus platyurus*) has a fold of skin along each side of its body.

**History of introduction and current distribution:** *Hemidactylus turcicus* of Eurasia and Africa is the oldest colonizer of the hemidactyline geckos in Florida. Commercial trade via cargo ships was probably responsible for its initial colonization of

*Hemidactylus turcicus* Mediterranean Gecko. Photo by R.D. Bartlett.

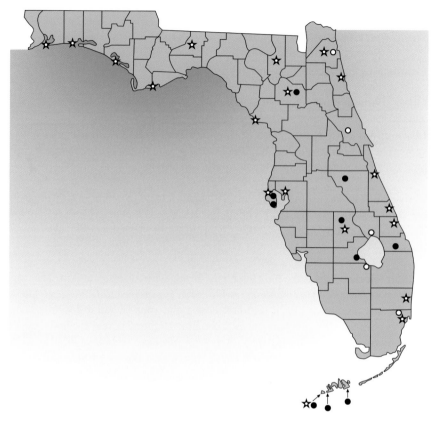

*Hemidactylus turcicus*

Key West, Monroe County, near the turn of the 20th century (Fowler, 1915; Stejneger, 1922), after which it quickly expanded its range to Miami, Dade County (Barbour, 1936). The dispersal pattern of *H. turcicus* in Florida shares the temporal and
spatial discontinuity associated with incidental human-assisted dispersal: Monroe
(Fowler, 1915), Dade (Barbour, 1936), Alachua (King, 1958), Hillsborough (Brown
and Hickman, 1970), Pinellas (McCoy, 1971), Duval (Meylan, 1977b), St. Lucie,
Indian River (Myers, 1978b), Broward (Wilson and Porras, 1983), Leon (Means,
1990b), Bay (Nelson and Carey, 1993), St. Johns (Wise, 1993), Escambia (Nelson
and Carey, 1993; Jensen, 1995), Highlands (Meshaka, 1995), Okaloosa (Jensen,
1995), Franklin (Means, 1996c, Collins and Irwin, 2001), Brevard (Criscione et al.,
1998), Levy (Means, 1999), and Columbia (Townsend and Reppas, 2001). We have
also found it in the counties of Hendry, Glades, Martin, Okeechobee, and Osceola
and have reports from Volusia. This discontinuous pattern of geographic expansion
in Florida has also been observed in other areas of its range (Davis, 1974).

Over the past several years, this gecko has disappeared from much of its Florida
range. On the Florida Keys, *H. turcicus* now appears to be practically absent. Re-

cent attempts to find it on Biscayne, Summerland, Vaca, or Big Pine Keys have been unsuccessful, whereas in the mid-1980s it was common (Frankenberg, 1984). Based on recent attempts to collect this species, populations of *H. turcicus* appear to have declined in southern and central peninsular Florida as well.

The apparent disappearance of this gecko from some areas has occurred following colonization by other hemidactyline geckos. On Key West, *H. mabouia, H. garnotii,* and *H. frenatus* have replaced *H. turcicus* (Meshaka et al., 1994a, 1994b). On other keys, *H. mabouia* alone appears to have replaced this gecko. For example, from 1992 through 1997, *H. mabouia* numbers increased while *H. turcicus* have apparently disappeared at sites on Sugarloaf Key and Key Largo.

This species appears to have been replaced by *H. garnotii* and *H. mabouia* in southern mainland Florida (Meshaka et al., 1994a), and by *H. garnotii* in central Florida (Meshaka, 1995). At several sites in St. Petersburg, Pinellas County, *H. garnotii* replaced *H. turcicus* within 4 years of the first discovery of *H. garnotii* in 1993. *Hemidactylus turcicus* had been established in St. Petersburg since the 1960s.

**Habitat and habits:** *Hemidactylus turcicus* occurs exclusively on buildings and similar human structures (Carr, 1940; Duellman and Schwartz, 1958; King, 1958; Nelson and Carey, 1993; Meshaka, 1995). It avoids trees and bushes. Far more individuals can be found on lighted buildings than unlighted ones, and incandescent lights attract more geckos and insects than do fluorescent lights and sodium vapor lamps (Nelson and Carey, 1993). Where no similar species occur, this gecko can be very numerous, with individuals foraging close to one another and moving about very little.

*Hemidactylus turcicus* is active at night, and it may emerge at dusk. Frankenburg (1984) noted that emergence of *H. turcicus* from its retreats coincides with darkness and ends by midnight. King (1958) observed a gradual decline in activity after a spike during 1900–2000 hrs. It is easily approachable and lacks the speed and wariness of other hemidactyline geckos in Florida. When pursued, individuals generally crawl slowly into their retreats or out of reach of their pursuer. Activity is continuous in Lake Placid but noticeably depressed from the end of November through February. Males defend sites and most social interactions occur in advance of foraging (Frankenberg, 1982b).

Having first colonized the Lower Keys in the late 1800s and having dominated buildings through the mid-1900s, *H. turcicus* is now all but absent. Likewise, this gecko appears to have been replaced by similar species in much of southern Florida, but the mechanisms of these replacements have not been identified. However, *H. turcicus* has characteristics that may put it at a disadvantage when compared to similar geckos. This gecko has activity patterns similar to those of *H. garnotii* but is socially dominated by *H. garnotii* (Frankenburg, 1984). This gecko has a restricted reproductive season and produces fewer clutches of eggs per year than *H. garnotii* (Meshaka, 1994b), *H. mabouia* (Meshaka et al., 1994d), and *H. frenatus* (Meshaka et al., 1994b). And finally, this gecko does not occupy vegetation. This species, however, does not originate from the tropics and could possibly escape its competitors at the northern edge of its Florida range.

**Table 16.** Diet of 9 *Hemidactylus turcicus* from Florida. Numbers indicate number of prey (number of lizards consuming each prey category).

| Taxa of Food Items | Males (N=5) | Females (N=4) | Total |
|---|---|---|---|
| Diptera | 48(2) | 99(2) | 147(4) |
| Dictyoptera | 2(2) | 1(1) | 3(3) |
| Lepidoptera | 5(4) | 5(3) | 10(7) |
| Coleoptera | 2(2) | 2(2) | 4(4) |
| Hymenoptera (Formicidae) | 1(1) | 0(0) | 1(1) |
| Homoptera | 0(0) | 1(1) | 1(1) |

**Reproduction:** Adult males (49.0 ± 4.3-mm SVL; range = 39.4–57.9; N = 75) and females (50.7 ± 4.3-mm SVL; range = 43.3–57.0; N = 66) are similar in body length (Meshaka, 1995). We report copulation in March in Lake Placid. King (1958) reported copulation in July in Gainesville, Alachua County. Multiple clutches of no more than two eggs are laid from April to August in Gainesville (King, 1958) and May to August in south-central Florida (Meshaka, 1995). Up to five clutches can be produced per year, with shelled eggs ranging from 8.4 to 10.6-mm in diameter (Meshaka, 1995).

**Diet:** This gecko is an insectivore, and both sexes feed on hard- and soft-bodied prey (Table 16).

**Predators:** We have no predator records of this species in Florida; however, the Cuban treefrog (*Osteopilus septentrionalis*), a predator of geckos (Meshaka, 2001), is a likely predator of *H. turcicus*.

*Pachydactylus bibronii* (Smith, 1845)

**Common name:** Bibron's Thick-Toed Gecko

**Other common name:** Bibron's Gecko

**Description:** The body color consists of a mottling of browns and black, and a dark stripe runs through the eye. The body is warty.

**Body size:** Adults from Florida are known to reach approximately 140-mm TL (Bartlett and Bartlett, 1999).

*Pachydactylus bibronii* Bibron's Thick-Toed Gecko. Photo by R.D. Bartlett.

**Similar species:** This gecko is unlikely to be mistaken for any other in Florida. The most similar, the ringed wall gecko (*Tarentola annularis*), has two white spots on each shoulder.

**History of introduction and current distribution:** A colony of this South African gecko has existed on buildings of urban areas of Bradenton, Manatee County, since the 1970s. Bartlett and Bartlett (1999) believe that the Manatee County colony is the result of a deliberate introduction of this popular pet trade species.

**Habitat and habits:** *Pachydactylus bibronii* occurs on buildings and is most active on warm summer nights.

**Reproduction:** This species is an egg layer; however, we do not have information on reproduction of Florida populations. In its native range, this species lays clutches of two eggs in rock crevices (Razzetti and Msuya, 2002).

*Pachydactylus bibronii*

**Diet:** We have not surveyed food habits for Florida populations. In its native range, this lizard eats insects, especially ants (Hymenoptera), termites (Isoptera), grasshoppers (Orthoptera), beetles (Coleptera), and flies (Diptera) (Razzetti and Msuya, 2002).

**Predators:** We know of no reported predators for this species in Florida populations.

---

## *Sphaerodactylus argus* Gosse, 1850

**Common name:** Ocellated Gecko

**Other common name:** none

**Description:** The body is brown, and the tail is rust or brown. Light-colored spots occur on the head and neck. The dorsal scales are smaller than those on the sides of the body and both are strongly keeled.

**Body size:** We have not measured specimens of this species in Florida, but adults are known to reach approximately 60-mm TL (Conant and Collins, 1998).

*Sphaerodactylus argus* Ocellated Gecko. Photo by R.W. Van Devender.

**Similar species:** The native Florida reef gecko (*Sphaerodactylus notatus notatus*) has large keeled dorsal scales. The ashy gecko (*Sphaerodactylus elegans*) lacks the light spots on the head and neck and its scales are small and granular.

**History of introduction and current distribution:** *Sphaerodactylus argus* is a Jamaican species first recorded from Florida in Key West, Monroe County (Savage, 1954). Its introduction was likely associated with commerce between Key West and the West Indies. Duellman and Schwartz (1958) could not find this gecko and thought that it no longer existed on Key West. Populations of this gecko were later reported from Key West (King and Krakauer, 1966; Love, 1978) and Stock Island, Monroe County (Wilson and Porras, 1983). Our surveys throughout the 1990s and those by Lawson and colleagues (1991) failed to locate this species anywhere on the Lower Florida Keys.

**Habitat and habits:** From scant records and observations, this species does not appear to have ever been numerous on the Lower Keys (Lazell, 1989).

*Sphaerodactylus argus*

**Reproduction:** The reproductive cycle of this egg laying species was suspected by Lazell (1989) to be the same as that of *S. elegans*.

**Diet:** We have not surveyed food habits of this species in Florida.

**Predators:** We know of no predator records for *S. argus* in Florida.

---

## *Sphaerodactylus elegans* MacCleay, 1834

**Common name:** Ashy Gecko

**Other common name:** MacCleay's Ashy Gecko

*Sphaerodactylus elegans* Ashy Gecko. Photo by W.E. Meshaka, Jr.

**Description:** The body color ranges from gray to light brown to gold. The head is marked with light stripes, and the snout is pointed. Juveniles have a banded pattern and a red tail. The scales are granular in this species.

**Body size:** The largest male and female, both 35-mm SVL, were from Key West, Monroe County.

**Similar species:** The native Florida reef gecko (*Sphaerodactylus notatus notatus*) has large keeled dorsal scales. The ocellated gecko (*Sphaerodactylus argus*) has light spots on its head and neck and its dorsal scales are strongly keeled.

**History of introduction and current distribution:** *Sphaerodactylus elegans* is native to Cuba. This gecko has been known to occur on Key West for more than 75 years (Stejneger, 1922). Carr (1940) thought that *S. el-*

Note: see text for localities in the Florida Keys

*Sphaerodactylus elegans*

*egans* was possibly native to Florida. Wilson and Porras (1983) categorize this species as a stowaway associated with commerce. This species has been found on Key West and Boca Chica Key (Duellman and Schwartz, 1958), Big Coppit Key (Wilson and Porras, 1983), Raccoon Key (Lawson et al, 1991), Boot Key (Epler, 1986), Summerland Key and Middle Torch Key (Lazell, 1989), all in Monroe County. We have found it on Key West, Stock Island, Cudjoe Key, and Big Pine Key. It appears that the growing ubiquity of the tropical gecko (*Hemidactylus mabouia*) in the early to mid-1990s was followed shortly thereafter by a precipitous drop in the number of *S. elegans* on the Keys. The causes of this negative association are unknown.

**Habits and habitat:** This gecko occurs on buildings and other human structures and under logs and tree bark. Lazell (1989) suggested that the historically large numbers of this gecko on Key West might have been in response to construction of buildings, which provided this species with quality habitat.

On warm nights, this gecko has been observed around floodlights on the walls of buildings and on the unlighted portions of wooden fences lining residential areas. On Stock Island, Monroe County, we found individuals at night under the bark of Australian pine trees, and on Cudjoe and Big Pine Keys specimens were taken during the day from beneath the bark of rotting logs.

Since 1993, we have observed decreasing numbers of *S. elegans* on the Lower Florida Keys as numbers of *H. mabouia* have increased. For example, *S. elegans* and the Mediterranean gecko (*H. turcicus*) were observed in near equal numbers on a hotel on Sugarloaf Key in 1985 (R.D. Bartlett, pers. comm., 1997). In 1994, *S. elegans* was rare, *H. turcicus* was absent, and *H. mabouia* was present. By 1997, though, *H. mabouia* was found exclusively and in higher numbers than in 1994 (R.D. Bartlett, pers. comm., 1997). These observations suggest that *H. mabouia* negatively impacts populations of *S. elegans*. The extent and causes of this phenomenon are unknown.

**Reproduction:** Lazell (1989) stated that this gecko breeds in the summer. We have collected gravid females containing 1–2 eggs in March and September. On 31 March 1994, eggs of *S. elegans* were collected with those of *H. mabouia* from under rocks in a hammock on Stock Island. One of the *S. elegans* eggs hatched on 1 April and the hatchling measured 22-mm SVL (R.D. Bartlett, pers. comm., 1996). Dunson and Bramham (1981) reported eggs in communal nest sites and found that these eggs were highly resistant to desiccation.

**Diet:** A single female (32-mm SVL) collected from Key West in May had eaten seven ants (Hymenoptera) and two flies (Diptera). None of these food items was over 2 mm in length.

**Predators:** We know of no reported predators of this species in Florida.

## *Tarentola annularis* (Geoffroy, 1827)

**Common name:** Ringed Wall Gecko

**Other common names:** Twin-Spotted Gecko, White-Spotted Wall Gecko.

**Description:** The body is light brown or grayish brown with slightly darker cross-bands. Two light-colored spots are located on each shoulder.

**Body size:** The largest male (104-mm SVL) and the largest female (100-mm SVL) were from Homestead, Dade County.

**Similar species:** Bibron's thick-toed gecko (*Pachydactylus bibronii*) lacks the two light-colored spots on each shoulder.

**History of introduction and current distribution:** *Tarentola annularis,* a native of Africa, was intentionally introduced in 1990 onto the buildings of an exotic animal dealership in Homestead. Bartlett (1997) also found this species on a building that was formerly occupied by a pet dealer in Ft. Myers, Lee County. The

*Tarentola annularis* Ringed Wall Gecko. Photo by R.D. Bartlett.

*Tarentola annularis*

Florida Fish and Wildlife Conservation Commission (1999–2002) reports this species from a pet store in Tallahassee, Leon County.

**Habitat and habits:** *Tarentola annularis* occurs on buildings and similar human structures and appears to be primarily nocturnal in its activity.

**Reproduction:** The body sizes of two males collected from Homestead were 80 and 104-mm SVL. The average body size of eight females collected from Homestead was 86.9 ± 7.6-mm SVL (range = 73–97). A female (84-mm SVL) collected in June contained two enlarged follicles (4 and 5 mm). Females collected in July, September, and October contained up to five sets of ova no larger than 3 mm. Percent left testis lengths of two males collected in July were 5.5 and 6.4% of body length. Two juveniles (37 and 39-mm SVL) were collected in October.

**Diet:** *Tarentola annularis* feeds on a wide range of invertebrates, especially beetles (Coleoptera), but it also eats other geckos (Table 17).

**Table 17.** Diet of 12 *Tarentola annularis* from buildings in Homestead, Dade County. Numbers indicate number of prey (number of lizards consuming each prey category).

| Taxa of Food Items | Total |
|---|:---:|
| Arachnida (Aranea) | 1(1) |
| Isopoda | 2(1) |
| Isoptera | 27(1) |
| Coleoptera | 8(7) |
| Dermaptera | 1(1) |
| Dictyoptera | 3(3) |
| Diptera | 1(1) |
| Hemiptera | 2(2) |
| Hymenoptera | 40(3) |
| Lepidoptera | 3(2) |
| Trichoptera | 2(2) |
| Asian House gecko (*Cosymbotus platyurus*) | 1(1) |
| House Gecko (*Hemidactylus frenatus*) | 1(1) |

**Predators:** We have no records; however, the tokay gecko (*Gekko gecko*) occurs with *T. annularis* and is a known predator of geckos.

---

# FAMILY TEIIDAE

## *Ameiva ameiva* (Linnaeus, 1758)

**Common name:** Giant Ameiva

**Other common name:** Jungle Runner

**Description:** The body color is greenish or bluish. The belly has 10 or 12 rows of enlarged, rectangular scales. The venter of males is blue, and is white in females.

*Ameiva ameiva* Giant Ameiva. Photo by R.D. Bartlett.

**Body size:** Adults reach 635-mm TL (Conant and Collins, 1998). The largest male (200-mm SVL) and female (159-mm SVL) were collected from Cape Florida State Recreation Area, Key Biscayne, Dade County.

**Similar species:** The native six-lined racerunner (*Cnemidophorus sexlineatus sexlineatus*) is much smaller than this species. Male *Ameiva ameiva* lack the strikingly blue or turquoise face of the much smaller rainbow whiptail (*C. lemniscatus*), and the giant whiptail (*C. motaguae*) has eight rows of enlarged, rectangular belly scales whereas *Ameiva ameiva* has 10–12.

**History of introduction and current distribution:** *Ameiva ameiva* is a polytypic species native to tropical South America. It was first reported from an undisclosed location in Florida in 1957 (Neill 1957), but was established in Miami, Dade County, as early as 1954 (Duellman and Schwartz, 1958). Separate colonies of two subspecies (*A. a. ameiva* and *A. a. petersi*) were reported from the Miami area (King and Krakauer, 1966). In 1992, we could not find *A. a. ameiva* at a reported Suniland site in Kendall, Dade County. The population of *A. ameiva* on Key Biscayne reported by Wilson and Porras (1983) is still in existence. That population was derived from releases of lizards from a zoological park exhibit, whereas the South Miami and Hialeah colonies reported by King and Krakauer (1966) were derived from releases associated with pet dealers or pet owners (Wilson and Porras, 1983). Bartlett and Bartlett (1999) state that the pet trade is the source of introduction of this species into Florida. We did not visit either the confirmed or unconfirmed northwest Miami populations of *A. ameiva* (Wilson and Porras, 1983). The taxonomic status of *A. ameiva* in Florida remains in question, and could be the product of intergradation between the two subspecies (Wilson and Porras, 1983). With the exception of the Suniland population, lizards of all colonies were predominantly blue in color and large in size (Wilson and Porras, 1983). We never saw individuals at Suniland, but all others were likewise bluish and large in body size, traits associated with the nominate form.

*Ameiva ameiva*

**Habitat and habits:** This lizard occupies open areas with nearby cover such as shrubs, boards, rocks, and woodpiles. Burrows have been found under concrete slabs and large rocks.

*Ameiva ameiva* actively forages during the day. Activity peaks in midmorning and is followed by a short second activity period in late afternoon. Individuals are usually observed moving in and out of ground vegetation in areas of filtered or direct sunlight, much in the way as the native *C. s. sexlineatus*. As it forages, it frequently digs with its forelimbs and pokes its snout under debris. On several occasions, we have observed male and female pairs foraging together. Large males occasionally forage in open areas. Females and juveniles generally remain closer to ground cover.

**Reproduction:** Bartlett and Bartlett (1999) report October breeding. We captured and measured two hatchlings (48.6 and 49.2-mm SVL) from Key Biscayne on 7 May 1997. Both hatchlings had prominent umbilical scars.

**Diet:** The stomach contents of the two aforementioned adult specimens found contained phasmids, roaches (Dictyoptera), beetles (Coleoptera), and various hymenopterans, larval insects (Lepidoptera), eggs of the brown anole (*Anolis sagrei*), and land snails (Gastropoda).

**Predators:** We know of no reports of predation on this species in Florida.

## *Cnemidophorus lemniscatus* (Linnaeus, 1758)

**Common name:** Rainbow Whiptail

**Other common name:** Rainbow Lizard

**Description:** The body of adult males has a brown middorsal stripe. The sides of the body are green, greenish yellow, or bright yellow with lighter spots. The tail is blue or bluish green. Coloration is brighter in adult males than in females or immature males. The sides of the head, throat, and anterior surfaces of the limbs of adult males are bright blue or turquoise. The sides of the head in females tend toward orange, they have 7–9 light, longitudinal stripes on the body, and the hind legs and tail are green.

**Body size:** The largest male (68.7-mm SVL) and the largest female (64.8-mm SVL) were collected in Miami, Dade County.

**Similar species:** The native six-lined racerunner (*Cnemidophorus sexlineatus sexlineatus*) has six light, longitudinal lines on the body and lacks the bright blue of male or orange of female *C. lemniscatus*. The giant ameiva (*Ameiva ameiva*) and giant whiptail (*C. motaguae*) are much larger and lack the bright blue or turquoise seen on the sides of the head and throat of males of this species.

*Cnemidophorus lemniscatus* Rainbow Whiptail. Photo by R.D. Bartlett.

**History of introduction and current distribution:** *Cnemidophorus lemniscatus,* native to Central and South America, was first reported from Hialeah, Dade County, in 1966 (King and Krakauer, 1966), where it no longer exists (Wilson and Porras, 1983). Currently, isolated populations exist in northern Dade County (Wilson and Porras, 1983; Bartlett, 1995b). *Cnemidophorus lemniscatus* is a favored pet trade species (Bartlett and Bartlett, 1999). The known colonies were derived from the pet trade (Wilson and Porras, 1983).

**Habitat and habits:** *Cnemidophorus lemniscatus* occupies the sandy soil of railroad rights-of-way and adjacent vegetation. Burrows are made in the soil and under objects such as concrete slabs and railroad ties.

*Cnemidophorus lemniscatus* is a diurnal, sun-loving species. It actively forages by moving slowly on the ground or above the ground in mats of vegetation. When threatened, this lizard quickly flees into thick vegetation, under objects, or into burrows.

*Cnemidophorus lemniscatus*

**Reproduction:** Conant and Collins (1998) suggested that *C. lemniscatus* found in Florida should be treated as a species complex because of the existence of at least four distinct forms, including bisexual and unisexual populations. In Florida, bisexual populations are confirmed, but it is unknown whether or not unisexual populations exist.

All adult females (62.2 ± 4.2-mm SVL) collected in early May (N = 4) and late July (N = 2) contained either vitellogenic follicles or oviductal eggs. Diameters of 10 follicles measured 5.9 ± 2.2 mm, and four oviductal eggs measured 17.8 ± 0.5 × 8.8 ± 0.5 mm. Average clutch size was 2.3 ± 0.5, based on the number of follicles and oviductal eggs. All females examined could produce at least two clutches of eggs each year.

A female captured in early May from Miami laid a clutch of two eggs on 19 May (P. Bedsole, pers. comm., 1997). These eggs were incubated at 30°C and hatched on 17 July. The hatchlings were 27 and 29-mm SVL. Three hatchlings, captured in early May, measured 29.9 ± 1.0-mm SVL. Left testis diameters of three adult males (65.3 ± 3.7-mm SVL) collected in early May measured 4.8 ± 0.2 × 3.8 ± 0.5 mm.

**Table 18.** Diet of 16 *Cnemidophorus lemniscatus* from Dade County. Numbers indicate number of prey (number of lizards consuming each prey category).

| Taxa of Food Items | Total |
|---|---|
| Arachnida | 4(3) |
| Coleoptera (adult) | 37(14) |
| Coleoptera (larvae) | 3(1) |
| Dictyoptera | 7(6) |
| Diptera | 5(3) |
| Hemiptera | 9(6) |
| Homoptera | 2(2) |
| Hymenoptera (Formicidae) | 144(9) |
| Lepidoptera | 14(6) |
| Odonata | 1(1) |
| Orthoptera | 1(1) |

**Diet:** This species preys upon a wide range of invertebrates, especially beetles (Coleoptera) and ants (Hymenoptera) (Table 18). This species also eats vegetation (R.D. Bartlett, pers. comm., 1996).

**Predators:** Corn snakes (*Elaphe guttata*) prey upon *C. lemniscatus* (R. St. Pierre, pers. comm., 1995).

---

## *Cnemidophorus motaguae* Sackett, 1941

**Common name:** Giant Whiptail

**Other common name:** Central American Whiptail

**Description:** The dorsum is light brown with yellow spots, and the sides are light gray with lighter spots. The venter is blue with black blotches. The tail is brownish blue at the base and becomes reddish toward the tip. The belly has eight rows of rectangular scales.

*Cnemidophorus motaguae* Giant Whiptail. Photo by W.E. Meshaka, Jr.

**Body size:** The largest male (136.6-mm SVL) and the largest female (122.5-mm SVL) were from Kendall, Dade County.

**Similar species:** The native six-lined racerunner (*Cnemidophorus sexlineatus sexlineatus*) is much smaller than this species and has longitudinal stripes. The exotic rainbow whiptail (*C. lemniscatus*) is also much smaller and male *C. motaguae* lack the strikingly blue face of male *C. lemniscatus*. The giant ameiva (*Ameiva ameiva*) has 10–12 rows of enlarged, rectangular belly scales whereas *C. motaguae* has only eight.

**History of introduction and current distribution:** *Cnemidophorus motaguae* is native to Central America. It was first reported from Kendall, Dade County (Bartlett, 1995b). We are also aware of a population in Opa-locka, Dade County (R. St. Pierre, pers. comm., 1995). The Kendall population has existed for at least eight years, and the Opa-locka colony has been in existence for over 20 years. The origin of the Kendall population is unknown, although presumed to be the result of a release or escape of pets. The Opa-locka population could have been founded by escaped individuals from a pet shop that formerly occupied the site.

**Habitat and habits:** *Cnemidophorus motaguae* occurs in open, sandy areas near water. The Kendall colony borders a canal, and the Opa-locka colony borders two artificial lakes. Burrows have been found in the soil or under objects such as concrete slabs and boards.

*Cnemidophorus motaguae* actively forages and basks on sunny days in open areas near cover such as bushes and burrows. It is rarely seen on overcast days. Male/female pairs have been observed foraging together. Daily activity usually peaks in midmorning. When threatened, it will quickly flee into thick vegetation, under objects, or into burrows.

**Reproduction:** Two females, collected in late July, contained three sets of previtellogenic follicles. The diameters of the largest set of previtellogenic follicles of two females (101 and 123-mm SVL) also collected in July were $3.7 \pm 0.6$ mm; N = 7 and $3.1 \pm 0.5$ mm; N = 13. Average left testis dimensions from three males (127.6, 134.7, and 136.6-mm SVL) collected in late July were $7.6 \pm 1.0 \times 5.8 \pm 0.3$ mm. A juvenile (37-mm SVL) was collected from Kendall in September.

**Diet:** This species is an insectivore, the diet of our sample having been comprised mostly of beetles (Coleoptera) and their larvae, roaches (Dictyoptera), and ants (Hymenoptera) (Table 19).

*Cnemidophorus motaguae*

**Predators:** We know of no reports of predators of this species in Florida.

**Table 19.** Diet of 20 *Cnemidophorus motaguae* from Kendall, Dade County. Numbers indicate number of prey (number of lizards consuming each prey category).

| Taxa of Food Items | Total |
|---|---|
| Coleoptera | 41(14) |
| Coleoptera larvae | 97(6) |
| Dictyoptera | 28(11) |
| Hymenoptera (Formicidae) | 43(8) |
| Lepidoptera | 13(6) |
| Arachnida: Aranea | 4(1) |

# FAMILY SCINCIDAE

## *Mabuya multifasciata* (Kuhl, 1820)

**Common name:** Brown Mabuya

**Other common names:** Bronze Skink, Sun Skink, Golden Skink

**Description:** The body color is drab olive. A bright yellow lateral stripe begins at the axilla and fades near the hind leg. The tail is yellow.

**Body size:** Adults are known to reach approximately 130-mm SVL. We have not measured specimens from Florida.

**Similar species:** This skink is found within the ranges of three native skinks: the peninsula mole skink (*Eumeces egregius onocrepis*), the southeastern five-lined skink (*E. inexpectatus*), and the ground skink (*Scincella lateralis*), none of which share the drab olive color and yellow tail of *Mabuya multifasciata*.

**History of introduction and current distribution:** *Mabuya multifasciata* is native to southeast Asia. This skink has occurred at the Kampong in Coconut Grove,

*Mabuya multifasciata* Brown Mabuya. Photo by T. Lodge.

Dade County, since at least 1990, and most noticeably since Hurricane Andrew (Meshaka, 1999b). It has also been observed in the vicinity of a reptile dealership in Ft. Myers, Lee County (R.D. Bartlett, pers. comm., 1997); however, its status in Ft. Myers is unknown. In light of its ubiquity in the pet trade, *M. multifasciata* is thought to have been introduced into Florida through the pet trade (Meshaka, 1999b).

**Habitat and habits:** This skink occurs in a lush tropical garden on the grounds of the Kampong, a private residence in Coconut Grove. It is terrestrial and semifossorial. It has been observed basking in open sunlight and foraging in and on mulch piles (Meshaka, 1999b).

**Reproduction:** *Mabuya multifasciata* is a livebearer. In Coconut Grove, pairs court in the spring (Meshaka, 1999b).

**Diet:** A large individual at the Kampong was observed wrestling a scorpion, apparently attempting to consume it. In captivity, *M. multifasciata* is an insectivore, although it will also eat ripe fruit and large individuals are capable of consuming small vertebrates.

*Mabuya multifasciata*

**Predators:** We know of no information regarding predation on this species in Florida; however, the eastern racer (*Coluber constrictor*) is present at the Coconut Grove site and a potential predator of this species.

# SQUAMATA: SERPENTES—SNAKES

## FAMILY TYPHLOPIDAE

### *Ramphotyphlops braminus* (Daudin, 1803)

**Common name:** Brahminy Blind Snake

**Other common name:** none

**Description:** The body is grayish black. The head is round and has the same thickness as the neck. The tail is short and pointed. This snake superficially resembles a thin, grayish black earthworm.

**Body size:** The largest individual (173-mm TL) was from Miami, Dade County

**Similar species:** This snake is unlikely to be confused with any other in Florida.

**History of introduction and current distribution:** *Ramphotyphlops braminus* has a pantropical distribution. It was first recorded from several disconnected sites

*Ramphotyphlops braminus* Brahminy Blind Snake. Photo by R.D. Bartlett.

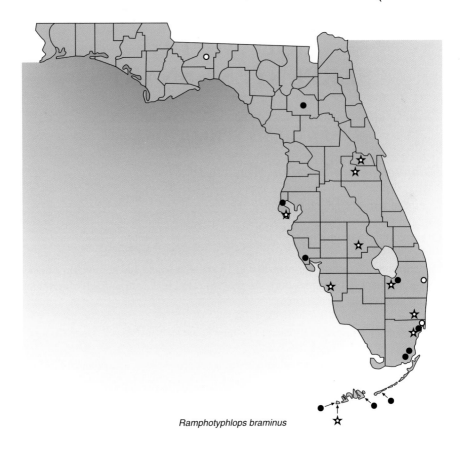

*Ramphotyphlops braminus*

in Dade County within a 2-year period (Wilson and Porras, 1983). It has since been recorded from the southeast shore of Lake Okeechobee, Palm Beach County (Delorey and Mushinsky, 1987); Key West, Big Pine Key, and Key Vaca in Monroe County (Ehrig, 1990; Watkins-Colwell and Watkins-Colwell, 1995c; Conant and Collins, 1998); Lee County (Conant and Collins, 1991); Clearwater, Pinellas County (Crawford and Somma, 1993a); Lake Placid, Highlands County (Meshaka, 1994c); West Palm Beach, Palm Beach County (Conant and Collins, 1998); Casselberry, Seminole County (Owen et al., 1998); Davie, Broward County (Krysko et al., 2000); and Winter Park, Orange County (Ernst and Brown, 2000). The FLMNH lists specimens from Alachua and Sarasota counties among its holdings. The Florida Fish and Wildlife Conservation Commission (1999–2002) reports two road-killed specimens from Tallahassee, Leon County.

This species was probably introduced into Florida incidentally with cargo (Wilson and Porras, 1983). Its subsequent dispersal through Florida has been associated with the ornamental plant trade (Wilson and Porras, 1983; Meshaka, 1994c), a dispersal agent responsible for such extreme records as those from Massachusetts (Wallach et al., 1991).

**Habitat and habits:** This snake inhabits Brazilian pepper thickets, Australian pine stands, disturbed tropical hardwood hammocks, pinelands, and residential areas.

This is a strictly fossorial species found under logs, rocks, and trash. Individuals of this species can be found under rocks with the Florida carpenter ant (*Camponotus abdominalis floridanus*). Its presence in ant nests has also been reported elsewhere (Crawford and Somma, 1993a).

**Reproduction:** *Ramphotyphlops braminus* is an all-female, parthenogenic species that lays eggs. The reproductive tracts of two individuals collected during early May 1997 from South Miami, Dade County, were examined. One individual (113.3-mm TL) contained two yolked follicles that measured 6.2 $\times$ 1.6 mm and 8.0 $\times$ 2.2 mm. The second individual (137.9-mm TL) contained six small (<1.0 mm) ovarian follicles.

**Diet:** We have not surveyed food habits of this species from Florida, but it is known to eat the eggs and pupae of ants and termites.

**Predators:** We observed a male Puerto Rican crested anole (*Anolis cristatellus*) (65-mm SVL) defecate an intact individual (90-mm TL). We also found this snake in fecal samples of the cane toad (*Bufo marinus*).

---

# FAMILY BOIDAE

## *Python molurus bivittatus* Kuhl, 1820

**Common name:** Burmese Python

**Other common name:** none

**Description:** The body color consists of a heavy pattern of tan, brown, and white. The body is very thick. Like other members of the family Boidae, this species has vestigial hind limbs which take the form of small hooks lateral to the vent.

**Body size:** The largest male (2880-mm SVL) and the largest female (2120-mm SVL) were from Everglades National Park, Monroe County.

**Similar species:** This snake is unlikely to be confused with any other in Florida. No other snake native to or established in Florida has vestigial hind limbs.

*Python molurus bivittatus* Burmese Python. Photo by R.D. Bartlett.

**History of introduction and current distribution:** *Python molurus bivittatus* is native to southeast Asia. Reports of this species in southern Florida extend back to the 1980s. Feral individuals, escaped captives, and intentional releases are regularly reported in urban Miami, Dade County, and elsewhere in Florida. Records of individuals exist at two sites; the saline glades near Flamingo, Monroe County, and Long Pine Key, Dade County, where a steady stream of reports have existed for more than a decade (Meshaka et al., 2000). Reports of this species also exist for Old Cutler Bay, Dade County.

Multiple observations of individuals of different size-classes supports the establishment of *P. m. bivittatus* in the mangrove forest of Everglades National Park (Meshaka et al., 2000). Four voucher specimens and scattered reports exist for *P. m. bivittatus* within a 2-mile stretch of Main Park Road, just north of Flamingo. Interestingly, two of the specimens, a female (2120-mm SVL) and a male (640-mm SVL), were collected on the road within 10 minutes of each other on 12 December 1995, following a cold front. A road-killed male of approximately 2000-mm TL (too damaged to measure) was also collected in the vicinity. A large unsexed individual was also collected in that vicinity during winter 2000 (W. Lofutus, pers. comm., 2002), and four other reports during that dry season have been submitted to the park (S. Snow, pers. comm., 2001). In September 2002, an individual (ca. 5000-mm TL) was killed on Long Pine Key, Hole in the Donut.

One of the specimens, a road-killed male (640-mm SVL), contained the remains of a gray squirrel (*Sciurus carolinensis*) further suggesting that persistence is possible in southern Florida. Other records from Everglades National Park, however, represent discarded pets; these records are of single individuals observed near buildings and picnic areas, without prey items in their digestive tracts. These specimens were also fatty, tame, and without scars (Meshaka et al., 2000). Several individuals have been observed from Old Cutler Bay, Dade County; however, too little information is available to determine if these individuals represent an established population. Thus, although established in the southern coastal fringe of Everglades National Park, the extent to which this species has colonized the mangrove fringe of extreme southern Florida remains unknown.

*Python molurus*

**Habitat and habits:** In Everglades National Park, *P. m. bivittatus* has been observed in a mangrove forest. Most observations have been of snakes basking in trees or crossing the road during cool days immediately following cold fronts.

**Reproduction:** The left testis of a male (2880-mm SVL) collected from Everglades National Park on 3 November 1997 measured 310 × 30 mm. We have no other information on reproduction for Florida populations. This species is oviparous and females are known to attend the clutch of eggs and warm them slightly with rhythmic contractions of body muscles.

**Diet:** The only diet information we have for this species in Florida is a *Sciurus carolinensis* found in the digestive tract of a road-killed male near West Lake, Everglades National Park. Rats (*Rattus*) and raccoons (*Procyon lotor*) in that habitat could provide a suitable food base for this species in the saline glades.

**Predators:** We know of no reports of predation on this species in Florida populations.

# CROCODYLIA: CROCODILIANS

## FAMILY ALLIGATORIDAE

### *Caiman crocodilus* (Linnaeus, 1758)

**Common name:** Spectacled Caiman

**Other common name:** none

**Description:** The body is brownish gray with darker blotches. A U-shaped bony protuberance (spectacle) occurs between the eyes.

**Body size:** Individuals in Florida rarely exceed 2000-mm TL. This species is known to reach approximately 2400-mm TL in tropical regions and we observed one individual very near this size in a canal near Florida City, Dade County.

**Similar species:** The U-shaped bony protuberance between the eyes is lacking in the American alligator (*Alligator mississippiensis*) and the American crocodile (*Crocodylus acutus*).

*Caiman crocodilus* Spectacled Caiman. Photo by S.L. Collins.

**History of introduction and current distribution:** *Caiman crocodilus* is native to Central America and northern South America. This species has long been popular in the pet trade and has been known to inhabit canals in Dade County since the late 1950s (Wilson and Porras, 1983). Citing information dating back to 1960, Ellis (1980) was the first to report that the species had become established as a breeding population. There is a persistent population in canals of the Homestead-Florida City area. Bartlett (1994) observed adult *C. crocodilus* at a site in Palm Beach County but could not determine if the species was established there. A clutch of 20 eggs collected in Brevard County is listed among the holdings of the FLMNH, but we know of no other specimens from that location. The Florida Fish and Wildlife Conservation Commission (1999–2002) reports reliable observations in Broward County. Ellis (1980) reported the species from Lake Jessup in Seminole County, but did not report breeding there. An adult was observed at Taylor Slough, Everglades National Park, for a few weeks in the 1980s, but may have been a lone individual that had dispersed into the park along the canal system. This species has not since been reported in Everglades National Park despite intensive fieldwork associated with *A. mississippiensis*.

*Caiman crocodilus*

**Habitat and habits:** *Caiman crocodilus* occurs in canals in urban and agricultural areas. We have observed it in small, weedy canals, and often in the entrances of culverts in larger canals. *Caiman crocodilus* is generally wary of humans. When disturbed, they often dive under the water and take refuge among submergent vegetation. Population levels have apparently been on the rise in some areas in recent years (R. St. Pierre, pers. comm., 1995).

**Reproduction:** This species is known to lay eggs, and nest-guarding behavior has been documented; however, we have no information on its reproductive cycle in Florida.

**Diet:** We have not surveyed food habits of this species from Florida. Ellis (1980) reported it eats fish, amphibians, and mammals in Florida.

**Predators:** We know of no reports of predation on this species in Florida.

# SPECIES OF UNCERTAIN STATUS

Unlike the species presented in previous accounts, the following 19 species have been reported in Florida but do not meet our criteria for establishment. See the introduction to the species accounts section for these criteria.

## *Pelodryas (Litoria) caerulea* (White, 1790)

**Common names:** White's Treefrog, Australian Green Treefrog, Dumpy Frog

A population that originated from a trader of exotic animals existed in Ft. Myers, Lee County, since at least 1990 (R. D. Bartlett, pers. comm., 1994). We also have reports of wild caught specimens from Sarasota, Sarasota County; however, we do not know if these specimens represent an established colony. Recent attempts to locate specimens have failed.

## *Pelusios subniger* (Lacépède, 1788)

**Common name:** East African Black Mud Turtle

We were informed that this Old World pelomedusid turtle has been established for more than 20 years in a rock pit in Dade County and that the colony represents an intentional introduction by a pet dealer. We have made three collecting trips to this site and have found no specimens.

## *Geochelone carbonaria* (Spix, 1824)

**Common name:** Redfoot Tortoise

This tortoise has been reported from a disturbed tropical hardwood hammock near Cutler Bay, Dade County, since the 1980s. We have no firsthand reports or records to validate its presence at the site.

## *Laudakia (Stellio) stellio* (Linnaeus, 1758)

**Common name:** Roughtail Rock Agama

123

Populations of this rock-dwelling lizard have been recorded from rock piles on the premises of a former exotic animal dealership in Homestead, Dade County, and from the parking lot of an active exotic animal dealership in Miami, Dade County.

## *Leiocephalus personatus* (Cope, 1862)

**Common names:** Haitian Curlytail Lizard, Green-Legged Curlytail Lizard, Masked Curlytail Lizard

In 1990, R.D. Bartlett photographed this Hispaniolan curlytail lizard near the Miami International Airport. Our efforts have yielded no specimens of this species.

## *Anolis extremus (extrema)* (Garman, 1887)

**Common names:** Extreme Anole, Barbados Anole

Bartlett and Bartlett (1999) reported this species from the Ft. Myers, Lee County area. Fewer than six specimens were observed in a 3-year period.

## *Anolis ferreus* (Cope, 1864)

**Common names:** Morne Constant Anole, Marie Gallant Sail-Tailed Anole

Bartlett and Bartlett (1999) reported a population of this lizard in the early 1990s near a golf course in Ft. Myers, Lee County.

## *Basiliscus plumifrons* Cope, 1876

**Common names:** Green Basilisk, Plumed Basilisk

We have been told that this species has occasionally been captured during the 1990s from the banks of canals in Dade County.

## *Gehyra mutilata* (Wiegmann, 1834)

**Common names:** Stump-Toed Dtella, Stump-Toed Gecko

A single *G. mutilata* was collected in July 1996 from a site previously occupied by an exotic animal dealership in Homestead, Dade County. An individual was also collected from a building in Gainesville, Alachua County, in February 1997.

## *Lepidodactylus lugubris* (Duméril and Bibron, 1836)

**Common name:** Mourning Gecko

This species has been observed on buildings of pet dealerships in Ft. Myers, Lee County, and Miami, Dade County. No individuals were observed at the Ft. Myers site after the pet dealership went out of business.

## *Phelsuma madagascariensis grandis* Gray, 1870

**Common name:** Madagascar Day Gecko

This diurnal gecko has been observed on trees within a city block of a pet dealership in Ft. Myers, Lee County. Individuals have also been reported from a residence near a pet dealership in Broward County. A deliberate release of this species at a residence in Homestead, Dade County, in 1990 is known to have failed. The species has also been reported from the Florida Keys, but we do not know the exact location(s). The FLMNH lists 19 specimens of this species, collected from Monroe County, among its holdings.

## *Ptyodactylus hasselquistii* (Donndorff, 1798)

**Common name:** Yellow Fan-Fingered Gecko

Individuals of this species were observed on a warehouse formerly occupied by a pet dealership in Gainesville, Alachua County, in 1995. Subsequent visits to the site have yielded no additional specimens.

## *Tarentola mauritanica* (Linnaeus, 1758)

**Common names:** Common Wall Gecko, Moorish Wall Gecko

Individuals of this African gecko have been reported from buildings in Ft. Myers, Lee County, and Miami, Dade County. The FLMNH lists a single specimen from Lee County among its holdings.

## *Tupinambis teguixin* (Linnaeus, 1758)

**Common names:** Black Tegu, Gold Tegu

This species has been observed in Everglades National Park (Butterfield et al., 1997), at a park in North Miami, Dade County, and at a park near Key Biscayne, Dade County.

## *Varanus niloticus* (Linnaeus, 1758)

**Common name:** Nile Monitor

Paul Moler informed us that this species has been observed repeatedly in a few locations in central Florida.

## *Acrochordus javanicus* Hornstedt, 1787

**Common name:** Java File Snake

Individuals of up to 2000-mm TL have been reported from a rock pit in Dade County. We have made three trips to this site and have found no specimens.

## *Boa constrictor* Linnaeus, 1758

**Common names:** Boa Constrictor, Red-Tailed Boa

This species has been reported since the mid-1980s from a pineland near Cutler Bay, Dade County. Individuals of different size-classes have been killed in fires at the site. In 1990, a 2500-mm TL female was captured at the site. Additionally, a road-killed male (950-mm SVL) was collected on 13 June 1981 in Lake Placid, Highlands County. The FLMNH lists a specimen from Dade County among its holdings.

## *Python reticulatus* (Schneider, 1801)

**Common name:** Reticulate Python

Large individuals of this species have been sighted several times in the mangrove forest of the saline glades just north of Flamingo, Everglades National Park, Monroe County. Most sightings were during the winter.

## *Python sebae* (Gmelin, 1789)

**Common name:** African Rock Python

This species has been observed sporadically near the Miami International Airport, Dade County.

# AFTERWORD
## The Human Role in the Colonization of Species in Florida

### Walter E. Meshaka, Jr.

Colonization is essentially a two-step process. The species first must disperse and then once there, it must become established. Ecological correlates of successful colonization are associated with both steps of the colonization process (Drake et al., 1989). Circumstances surrounding historic and, especially, present-day Florida have been amenable to both steps in the colonizations of nonindigenous species. Primarily through commerce, incidental or deliberate human-mediated dispersal was responsible for the arrival of all of the 40 nonindigenous amphibian and reptile species to Florida, which are found overwhelmingly through southern Florida. Dispersal almost always begins along the coast. This finding corroborates the pattern detected in earlier treatments of 36 (Butterfield et al., 1997), 25 (Wilson and Porras, 1983), and 13 (King and Krakauer, 1966) species. Therefore, it is not surprising that Florida's oldest and most widespread exotic species were native (e.g., brown anole) or exotic (e.g., Mediterranean gecko) stowaways from the West Indies, where they were transported to port cities in extreme southern Florida. In that regard, commercial trade with the lower Florida Keys is so old that dispersal of the greenhouse frog, Cuban treefrog, brown anole, ocellated gecko, and ashy gecko could just as easily have begun with aboriginal Indians and Spaniards as with later Americans, all of which attests to their close association with human-mediated dispersal (Carr, 1940; Duellman and Schwartz, 1958; Duellman and Crombie, 1970; Lazell, 1989; Butterfield, 1996; Meshaka, 1996a).

The literature, as summarized in this book, reveals that the recent spike of exotic amphibian and reptile introductions to Florida over the past 30 to 50 years has been primarily through the pet trade. Centers for the importation of these exotic species also tend to be coastal and primarily in the subtropical climate of southern Florida, a region that is the epicenter for other segments of Florida's exotic biota (Simberloff, 1997). Consequently, the centers of distribution for most of these exotic amphibians and reptiles continue to be southern Florida; the highest counts of exotic species are found in Dade and Broward counties (see Figure 2). In addition, through the pet trade, initial introductions can occur almost simultaneously at multiple sites distant from one another, as noted in some species accounts in this book. Ironically, most of the early incidental stowaway species, like the Cuban treefrog, brown anole, and Mediterranean gecko, have in the past 30 or so years also become part of the pet trade. Exceptionally, the single deliberate introduction unrelated to the pet trade—the cane toad, for pest control—did not establish until 20 years later through an escape of imported individuals destined for the pet trade.

If colonization success is to be measured by dispersal rates and persistence, then the most successful exotic species will exploit the many opportunities for dispersal. The high vagility that enabled the earliest stowaway species to disperse *to* Florida in ships is now replayed over and over on vehicles that carry them in cargo (building materials and plants, to name a few) or on the vehicle itself to new locations *within* Florida faster and more frequently than ever before. In some instances these species are even transported back to native areas in the West Indies.

The high vagility in the agency of humans that rewarded the early exotic amphibians and reptiles in Florida with broad geographic distribution is also the mark of the most successful of the recent successful colonizers, such as many of the hemidacytline geckos and the Brahminy blind snake. Some species, old (greenhouse frog) or new (Brahminy blind snake), continue to be dispersed almost exclusively as incidental stowaways with humans. Other species, like the tokay gecko, disperse almost as rapidly through their primary association with the pet trade as escapes or releases by both the retailer and the consumer. Still other species, like the Cuban treefrog, brown anole, and the hemidactyline geckos, have the best of both worlds, whereby individuals continue to disperse as stowaways *and also* as commerce in the pet trade.

Not surprisingly, no matter how vagile the species, colonization is a dead end for all but the species that can establish viable populations. As a collective ecological unit, most of the 40 exotic species of amphibians and reptiles in Florida are small and medium-sized insectivorous tropical lizards that can be found around various forms of human disturbance. As a subset of the 40 species, the most successful almost without exception have the following characteristics:

1. They are sexually mature within one year of life.
2. They produce multiple clutches during a prolonged breeding season.
3. They include a wide range of invertebrate prey in their diets (especially beetles, roaches, and ants).
4. They excel around human habitation.
5. Three highly successful species, the Cuban treefrog, brown anole, and greenhouse frog, thrive in various kinds of intact natural habitats, several of which occur in their native centers of distribution.

The advantages of the first two traits are that reproductive individuals are recruited into a population quickly and at almost any time of the year and their mixed generations protect the population from episodic catastrophes. These findings apply to the Cuban treefrog and are clearly applicable to those exotic species sensitive to drought or frost. A broad diet is another advantage to colonization (Ehrlich, 1989), providing the species with a range in prey types from which to exploit. In particular, beetles, ants, and roaches, which are favored among many of the insectivores, are common around disturbed sites, and beetles will come to lights.

The immediate advantage of being a human commensal is broad in scope. The various forms of human disturbance ranging from mildly altered habitats to cities create a protective environment for new colonizers. Such new habitats provide food in the form of light-attracted insects as well as other taxa, like spiders and roaches, that are differentially exploited by different species. Buildings also provide shelter, breeding sites, and protection from the elements.

Furthermore, disturbed sites in many forms often marginalize native species, including potential predators and competitors. The danger here is that native species could otherwise provide some level of resistance to colonization of exotic species. A look in most any backyard in southern Florida will verify the abundance of exotic species and the dearth of native species, including predators and potential competitors. Indeed, it appears that geckos, especially the hemidactylines, have only themselves with which to compete for food in urban settings. Anoles, the other taxonomic block of Florida colonizers, are undeterred by the single native anole species, and are for the most part niche-packing among themselves. Furthermore, native predators of lizards are simply rare if even present in urban and certain disturbed settings. The Cuban treefrog, cane toad, and northern curlytail lizard eat the species that come closest to being competitors. Predators of the cane toad are negligible, and for reasons yet to be determined the greenhouse frog is ubiquitous in practically whatever habitat it occurs.

Notwithstanding thermal limits, quite predictably then, these are the same species that can be expected to disperse most quickly northward into fractured systems from what is otherwise primarily a southern Florida phenomenon. This pattern is already reflected in the spatial discontinuity of colonization for many of these exotic species, which links the colonization of these species with Florida's urban development. The thermal advantage afforded to building-dwelling species will be extrapolated to inhabitants of urban heat islands, and the species that colonize natural habitats will have the further advantage of invading vast terrain that for other species is inhospitable. Such has been the case for the Cuban treefrog, and a good natural history study will, I am sure, reveal exactly the same pattern for the other tropical species.

Instant transportation and extensive human population expansion are causing an inadvertent experiment in biogeography at temporal and spatial scales never before experienced. The explosion of these nonindigenous species in Florida indicates a vital need for scientific research for a number of reasons. First, with humans present, there is opportunity for species colonization as is true for the other segments of the exotic biota of Florida. Second, the few studies that have examined ecological relationships of exotic species with other species provide evidence of or suggest potential for predator-prey or competitive relationships between exotic species and native species. Third, unlike Hawaii, Florida is physically connected to the North American continent. Consequently, the study of these species in Florida brings with it urgency as humans alter more habitat and provide entry for increasing numbers of exotic species. These species will interact with native species and some will disperse into other states. The colonization of the Mediterranean gecko and greenhouse frog in sites outside of Florida underscores this threat.

By no means originally, but wholeheartedly, I argue that ecological correlates of colonization success should be used as a starting point to explain why a nonindigenous species succeeds. Although too many confounding variables preclude reliable predictions of which species will definitely succeed, if well studied these correlates provide a strong measure of the likelihood of colonization success. Vigilant inventory of species and painstaking life history studies that without prejudice determine why a species is succeeding will identify the problems and perhaps provide the solutions to these questions: What is a given nonindigenous species doing to other species? How if at all can it be eradicated? How can its colonization be prevented elsewhere? and most difficult of all, Why did a species not succeed?

Because natural systems in Florida are ever more isolated, shrinking in size (Cox et al., 1994), and more distant to reach, many young Floridians do not know what has been lost or what continues to be lost each day. Backyard wildlife of any sort is a source of interest and wonder. High attendance at zoos and national parks, and a popular pet industry attest to the public's aesthetic interest in nature. To that end, daily sightings of exotic species in the backyard can become more exciting than the remote possibility of seeing a native species.

Nonetheless, although the overwhelming preponderance of exotics is in urbanized areas only, the position that an exotic species is better than nothing is still unjustifiable. No Florida system in this day and age is truly closed. Technology and human overpopulation increase the opportunity and therefore the likelihood of passive dispersal to natural areas, which in turn increase the likelihood of contact with natural systems and their components. Such a meeting, however unintentional, provides opportunity for an exotic to found a population, to compete with or prey upon native species, and to transmit a pathogen for which native species may or may not be resistant. Humans are not the only ones capable of arranging such a scenario between exotic and native species. Many avian predators, resident or migratory, flying between urban and natural areas, have the chance to eat an exotic and transmit a disease to native species.

Aggravating this problem is that natural systems change, and species adapt. Given time, some of these exotics thought to be building-only species invade seminatural and natural systems, as in the case of the Cuban treefrog, greenhouse frog, and brown anole, to name a few. The most dramatic example is that the Burmese python is established in the mangrove forests of extreme southern Florida. The presence of an exotic species of any taxon in nature could be easily overlooked if its population size is small, if the species is cryptic, if it resembles a native species, if no one is studying the species or system, or if the species has not yet been provided with the opportunity to colonize a natural habitat. On the positive side, some members of the native herpetofauna can at least persist in marginalized habitats. This is critical in light of rapidly expanding urban systems. If exotic species are decimating urban-dwelling natives species (e.g., native treefrogs versus the Cuban treefrog, southern toad versus the cane toad, the green anole versus exotic anoles), protecting these marginal native holdouts becomes ever more critical as the size and numbers of natural areas diminish.

Although truly nightmarish herpetological scenarios have not yet been documented in Florida at the single species level, the arrival of each new species collectively brings the state closer to losing this game of biological Russian roulette.

Cumulative negative impacts by exotic species chip away at the integrity of the system and accelerate ecocollapse. Cuban treefrog–native hylid and brown anole–green anole interactions speak to this point. For these reasons, arguments are compelling for preservation of *indigenous species and systems*. It is ultimately best to err on the side of safety so as to minimize the amount of regrets as the presence of exotic species incrementally, or in devastating fashion, fractures indigenous components and communities.

If we forget our role of stewardship, do not fund strong resource management, and allow unbridled development, these nonindigenous species will serve as a barometer of our folly. First, natural systems become fragmented and degraded, and next replaced with overpopulated human systems. We have now reached a point where we must either abandon all hope of natural system integrity and live with that abandonment or take serious steps to rectify the situation. What to do at this crossroads? Taken to its logical end, a future version of this book could detail a swarm of exotics into natural systems that will mutely proclaim total abandonment of stewardship of natural areas and their inhabitants. Another plausible scenario is an eventual absence of these mostly diminutive exotic species from their urban habitats, in its own way signifying a sterile lifeless human environment, one that has lost its connection with the natural world humans are obligated to protect. Doubtless, these extremes never dreamed of in earlier times are not only possible but also probable.

If the ecosystems of future Florida are to be a closer match with the past than with the present, the tide of exotic species can only be stemmed by the same human hand that has created the present milieu. Solutions, by no means novel, but all achievable, exist:

1. Implementation of a more responsible framework within which exotic species are imported, distributed, and maintained, in Florida. The state of Florida is quite clear on the regulation of foreign animals (Florida Statute 372.265). To summarize, a permit issued by the Florida Fish and Wildlife Conservation Commission (FWCC) is required for individuals to import for sale or use or to release in the state any nonindigenous animal species. Clarification regarding the exotic species threats would provide the FWCC with necessary information to make the best decisions, including those relating to captive breeding programs whose escapees could act as founders for new colonies. To be effective, this recommendation requires strict enforcement.

2. A clear-cut and strictly enforced statewide development plan that solves once and for all the amount of space to be allotted to nature, human development, and human recreation. The general ambiguity associated with Florida zoning adds to the difficulty of implementing this recommendation, but it should be understood that the larger the natural habitat parcel, the better the chance of its biotic community remaining intact.

3. An aggressive campaign to return natural areas to being natural areas combined with judicious purchases of contiguous tracts of natural habitat. Intact

systems decrease the likelihood of species invasion by providing various levels of predatory and competitive resistance. The purchase of large parcels by
the state of Florida has been a good step in this direction.

4. Incorporation of Florida ecology into the grade school and high school cur-
   ricula.
5. State funding for accountable studies on the biologies of these exotic species
   to better understand why they succeed and what can be done to exclude them
   from managed systems.

All of these recommendations *can* be implemented with strong justification, and
taken together they provide a way through a vexing conservation issue.

In my opinion, this book has done little if it hasn't both convincingly revealed
human responsibility for this exotic species list and revealed the importance of
knowing more about the lives of these species to deal more effectively with them.
The future direction of Florida should encompass the future of *all* of its resident
species, and we are capable of and responsible for actively directing that future.
Only then will we begin to understand and fulfill the proper role of steward and
not abuser of Florida's great legacy, its natural treasure.

# GLOSSARY

**Amplexus:** the clasping of a female frog or toad by a male during copulation.

**Axilla:** the region just behind the front leg; the armpit.

**Bufotoxin:** toxins secreted (especially from parotoid glands) by toads of the genus *Bufo.*

**Carapace:** the upper shell of a turtle.

**Cloaca:** the common chamber into which the digestive, urinary, and reproductive systems lead. The cloaca empties to the outside via the vent.

**Crest:** a ridge or series of spine-like scales along the midline of the neck, back, or tail.

**Dewlap:** a flap of skin on the mid-ventral surface of the throat and neck of some lizards. In some species (especially the anoles) this flap of skin is brightly colored, supported by cartilage, and can be extended or retracted. In these species, the dewlap is used as a signal in behaviors associated with mating and territorial defense.

**Dextral:** referring to the right side. A dextral vent tube opens toward the right side of the body.

**Dorsolateral:** intermediate between the dorsal (back) and lateral (side) surfaces.

**Dorsum:** the upper or "back" side of an animal.

**Endolymphatic sacs:** often found located on the ventrolateral surfaces of the neck of the Indo-Pacific gecko *Hemidactylus garnotii,* these structures appear as bags of chalky-white substance just under the skin. These structures serve as a means of storing calcium, important for egg shell production.

**Femoral pores:** a series of openings on the posterior edge of the ventral surface of the thighs of some lizards. These pores are the openings of exocrine glands and are usually better developed in males than females.

**Fossorial:** living under ground; burrowing species rarely if ever seen above ground.

**Gravid:** pregnant, carrying eggs or babies.

**Intergrade:** the result of breeding between two subspecies of the same species. Offspring usually show a mix of characters seen in each subspecies. Intergradation is normal where two subspecies naturally meet. It could occur unnaturally if one subspecies is introduced into the habitat occupied by a native subspecies.

**Interorbital crests:** hardened ridges located on the top of the head and between the eyes of some toads.

**Lamellae:** a series of overlapping plates or scales, such as those found on the underside surfaces of the toes of anoles and geckoes.

**Melanistic:** darker in color, due to the presence of large amounts of the pigment melanin.

**Mesophytic:** plants that are adapted to environments that are not extremely wet or extremely dry.

**Metachrosis:** a change in color. Some lizards, such as the anoles, can change color rapidly and dramatically. *Anolis equestris,* for example, can change from brilliant green to dark brown.

**Metamorph:** a newly transformed frog or toad. The tiny frog or toad may have transformed from a free-swimming tadpole (as most frogs and toads do) or may have emerged directly from an egg in those species that pass through the tadpole stage while still in the egg (like some frogs of the genus *Eleutherodactylus*).

**Nuchal:** on the dorsal surface of the neck.

**Pantropical:** found at tropical latitudes all around the world.

**Parotoid gland:** a large gland just behind the eye, extending onto the neck and sometimes onto the shoulder. Found on toads (genus *Bufo*).

**Philopatry:** refers to the tendency of an individual to stay in, or return to, its home area.

**Precloacal pores:** a series of openings on the scales just anterior to the vent of some lizards. These pores are the openings of exocrine glands.

**Rugose:** rough, covered with ridges.

**Semifossorial:** frequently found underground in burrows, but does not live there all the time.

**Sinistral:** referring to the left side. A sinistral spiracle is located on the left side of a tadpole's body.

**Spinose:** covered in spines or pointed protrusions. These may be hard or soft.

**Spiracle:** the opening from the gill chamber to the outside of the body of tadpoles.

**Terrestrial:** living on the ground.

**Tuberculate:** covered with tubercles or small, rounded bumps.

**Vagility:** the ability to be transported readily, naturally or otherwise. High vagility should increase the chance that an exotic will become established and/or increase its range.

**Vent:** the opening that connects the cloaca to the external environment.

**Venter/Ventral:** the lower or "belly" side of an animal.

**Vertebral:** on the midline of the back.

**Vestigial:** highly reduced, small.

**Xeric:** dry conditions, especially as related to soil.

# APPENDIX A
## Table of Contents with Higher Taxonomic Units in Alphabetical Order

CONTENTS

# APPENDIX B

## Cross Reference of Scientific and Common Names of Animal Species Mentioned in This Book

| Scientific name | Common name |
|---|---|
| *Acrochordus javanicus* | Java file snake |
| *Agama agama* | common agama |
| *Agkistrodon piscivorus* | cottonmouth |
| *Alligator mississippiensis* | American alligator |
| *Ameiva ameiva* | giant ameiva |
| *Anolis carolinensis* | green anole |
| *A. chlorocyanus* | Hispaniolan green anole |
| *A. cristatellus* | Puerto Rican crested anole |
| *A. cybotes* | largehead anole |
| *A. distichus* | bark anole |
| *A. equestris* | knight anole |
| *A. extremus (extrema)* | extreme anole |
| *A. ferreus* | morne constant anole |
| *A. garmani* | Jamaican giant anole |
| *A. porcatus* | Cuban green anole |
| *A. sagrei* | brown anole |
| *Basiliscus plumifrons* | green basilisk |
| *B. vittatus* | brown basilisk |
| *Boa constrictor* | boa constrictor |
| *Bufo marinus* | cane toad |
| *B. quercicus* | oak toad |
| *B. terrestris* | southern toad |
| *Buteo lineata* | red-shouldered hawk |
| *B. platypterus* | broad-winged hawk |
| *Bubulcus ibis* | cattle egret |
| *Caiman crocodilus* | spectacled caiman |
| *Calotes mystaceus* | Indochinese bloodsucker |
| *Camponotus floridanus* | Florida carpenter ant |
| *Canus familiarus* | domestic dog |
| *Clarias batrachus* | walking catfish |
| *Cnemidophorus lemniscatus* | rainbow whiptail |
| *C. motaguae* | giant whiptail |
| *C. sexlineatus sexlineatus* | six-lined racerunner |
| *Coluber constrictor* | eastern racer |
| *Corvus brachyrhynchos* | American crow |
| *Cosymbotus platyurus* | Asian house gecko |
| *Crocodylus acutus* | American crocodile |

| | |
|---|---|
| *Crotalus adamanteus* | eastern diamondback rattlesnake |
| *Ctenosaura pectinata* | Mexican spinytail iguana |
| *C. similis* | black spinytail iguana |
| *Cyanocitta cristata* | blue jay |
| *Diadophis punctatus* | ringneck snake |
| *Drymarchon corais couperi* | eastern indigo snake |
| *Elaphe guttata* | corn snake |
| *E. obsoleta quadrivittata* | yellow rat snake |
| *Eleutherodactylus coqui* | Puerto Rican coqui |
| *E. planirostris* | greenhouse frog |
| *Eumeces egregius onocrepis* | peninsula mole skink |
| *E. inexpectatus* | southeastern five-lined skink |
| *Eurycotus floridana* | Florida stinking roach |
| *Felis catus* | domestic cat |
| *Gastrophryne carolinensis* | eastern narrowmouth toad |
| *Gehyra mutilata* | stump-toed dtella |
| *Gekko gecko* | tokay gecko |
| *Geochelone carbonaria* | redfoot tortoise |
| *Gopherus polyphemus* | gopher tortoise |
| *Gonatodes albogularis fuscus* | yellowhead gecko |
| *Hemidactylus frenatus* | house gecko |
| *H. garnotii* | Indo-Pacific gecko |
| *H. mabouia* | tropical gecko |
| *H. turcicus* | Mediterranean gecko |
| *Heterodon platyrhinos* | eastern hognose snake |
| *Hyla chrysoscelis* | Cope's gray treefrog |
| *H. cinerea* | green treefrog |
| *H. gratiosa* | barking treefrog |
| *H. squirella* | squirrel treefrog |
| *Ictaluras natalis* | yellow bullhead catfish |
| *Iguana iguana* | green iguana |
| *Lanius ludovicianus* | loggerhead shrike |
| *Laudakia (Stellio) stellio* | roughtail rock agama |
| *Leiocephalus carinatus armouri* | northern curlytail lizard |
| *L. personatus* | Haitian curlytail lizard |
| *L. schreibersii* | red-sided curlytail lizard |
| *Lepidodactylus lugubris* | mourning gecko |
| *Mabuya multifasciata* | brown mabuya |
| *Mimus polyglottus* | northern Mockingbird |
| *Nerodia fasciata* | southern water snake |
| *Osteopilus septentrionalis* | Cuban treefrog |
| *Otus asio* | eastern screech owl |
| *Pachydactylus bibronii* | Bibron's thick-toed gecko |
| *Pelodryas (Litoria) caerulea* | White's treefrog |
| *Pelusios subniger* | East African black mud turtle |

| | |
|---|---|
| *Phelsuma madagascariensis grandis* | Madagascar day gecko |
| *Phrynosoma cornutum* | Texas horned lizard |
| *Pogonomyrmex badius* | Florida harvester ant |
| *Polioptila coerulea* | blue-gray gnatcatcher |
| *Procyon lotor* | raccoon |
| *Progne subis* | purple martin |
| *Pseudacris crucifer* | spring peeper |
| *Pseudacris nigrita verrucosa* | Florida chorus frog |
| *Pseudemys floridana peninsularis* | Peninsula cooter |
| *P. nelsoni* | Florida redbelly turtle |
| *Ptyodactylus hasselquistii* | yellow fan-fingered gecko |
| *Python molurus bivittatus* | Burmese python |
| *P. reticulatus* | reticulate python |
| *P. sebae* | African rock python |
| *Quixcalus major* | boat-tailed grackle |
| *Ramphotyphlops braminus* | Brahminy blind snake |
| *Rana catesbeiana* | bullfrog |
| *R. sphenocephala* | Florida leopard frog |
| *Rattus spp.* | rat |
| *Romelea microptera* | southeastern lubber grasshopper |
| *Sceloporus woodi* | Florida scrub lizard |
| *Scincella lateralis* | ground skink |
| *Sciurus carolinensis* | gray squirrel |
| *Sphaerodactylus argus* | ocellated gecko |
| *S. elegans* | ashy gecko |
| *S. notatus notatus* | Florida reef gecko |
| *Storeria dekayi* | brown snake |
| *Streptopelia decaocto* | Eurasian collared-dove |
| *Strix varia* | barred owl |
| *Tarentola annularis* | ringed wall gecko |
| *T. mauritanica* | common wall gecko |
| *Terrapene carolina* | eastern box turtle |
| *Thamnophis sauritus* | eastern ribbon snake |
| *T. s. sackenii* | peninsula ribbon snake |
| *T. sirtalis* | common garter snake |
| *T. s. sirtalis* | eastern garter snake |
| *Trachemys scripta elegans* | red-eared slider |
| *Tupinambis teguixin* | black tegu |
| *Tyto alba* | barn owl |
| *Varanus niloticus* | Nile monitor |
| *Zenaida asiatica* | white-winged dove |
| *Z. macroura* | mourning dove |

# REFERENCES

Achor, K.L. and P.E. Moler. 1982. Geographic distribution: *Anolis equestris* (knight anole). Herpetol. Rev. 13:131.

Allen, E.R. and W.T. Neill. 1955. Establishment of the Texas horned toad, *Phrynosoma cornutum*, in Florida. Copeia 1955:63–64.

Allen, E.R. and W.T. Neill. 1958. Giant toads from the tropics. Florida Wildlife 12:30–32, 42.

Allen, E.R. and R. Slatten. 1945. A herpetological collection from the vicinity of Key West, Florida. Herpetologica 3:25–26.

Ashton, R.E., Jr. 1976. County records of reptiles and amphibians in Florida. Florida State Mus. Herpetol. Newsletter 1:1–13.

Ashton, R.E., Jr. and P.S. Ashton. 1988a. Handbook of Reptiles and Amphibians of Florida Part One: The Snakes. 2nd ed. Windward Publishing. Miami, FL.

Ashton, R.E., Jr. and P.S. Ashton. 1988b. Handbook of Reptiles and Amphibians of Florida Part Three: The Amphibians. 2nd ed. Windward Publishing. Miami, FL.

Ashton, R.E., Jr. and P.S. Ashton. 1991. Handbook of Reptiles and Amphibians of Florida Part Two: Lizards, Turtles and Crocodilians. Windward Publishing. Miami, FL.

Austin, D.F. and A. Schwartz. 1975. Another exotic amphibian in Florida, *Eleutherodactylus coqui*. Copeia 1975:188.

Austin, S. 1975. Exotics. Florida Naturalist 48:2–5.

Babbitt, K.J. and W.E. Meshaka, Jr. 2000. Benefits of eating conspecifics: effects of background diet on survival, and metamorphosis in the Cuban treefrog (*Osteopilus septentrionalis*). Copeia 2000:469–474.

Bailey, J.W. 1928. A revision of the lizards of the genus *Ctenosaura*. Proc. U.S. Nat. Mus. 73:1-58.

Bancroft, G. T., J. S. Godley, D. T. Gross, N. N. Rojas, D. A. Sutphen and R. W. McDiarmid. 1983. Large-scale operations management test of use of the white amur for control of problem aquatic plants. The herpetofauna of Lake Conway: species accounts. Final report. Miscellaneous Paper A-83–5, U.S. Army Engineers Waterways Experiment Station, CE. Vicksburg, MS.

Barbour, T. 1910. *Eleutherodactylus ricordii* in Florida. Proc. Biol. Soc. Washington 23:100.

Barbour, T. 1931. Another introduced frog in North America. Copeia 1931:140.

Barbour, T. 1936. Two introduced lizards in Miami, Florida. Copeia 1936:113.

Bartlett, R.D. 1980. Non-native reptiles and amphibians in Florida. Indigo, News Bull. Florida Herpetol. Soc. 1: 52–54.

Bartlett, R.D. 1988. In Search of Reptiles and Amphibians. E.J. Brill. New York.

Bartlett, R.D. 1994. Florida's alien herps. Reptile and Amphibian Mag. March-April 1994: 56–73, 103–109.

Bartlett, R.D. 1995a. The anoles of the United States. Reptiles Mag. 2:48–62, 64–65.

Bartlett, R.D. 1995b. The teiids of the southeastern U.S. Trop. Fish Hobbyist 43:112, 114–119, 121–122, 124–126.

Bartlett, R.D. 1997. The geckos of Florida: native and alien. Reptilian 4:44–50.

Bartlett, R.D. and P.B. Bartlett. 1995. Geckos: Everything About Selection, Care, Nutrition, Diseases, Breeding, and Behavior. Barron's Educational Series. Hauppauge, NY.

Bartlett, R.D. and P.B. Bartlett. 1999. A Field Guide to Florida Reptiles and Amphibians. Gulf Publishing Co. Houston, TX.

Bell, L.N. 1953. Notes on three subspecies of the lizard *Anolis sagrei* in southern Florida. Copeia 1953:63.

Brach, V. 1976. Habits and food of *Anolis equestris* in Florida. Copeia 1976: 187–189.

Brach, V. 1977. Notes on the introduced population of *Anolis cristatellus* in south Florida. Copeia 1977:184–185.

Brown, L.N. 1972. Presence of the knight anole (*Anolis equestris*) on Elliott Key, Florida. Florida Nat. 45:130.

Brown, L.N. and G. C. Hickman. 1970. Occurrence of the Mediterranean gecko in the Tampa, Florida, area. Florida Nat. 43:68.

Butterfield, B.P. 1996. Patterns and processes of invasions by amphibians and reptiles into the West Indies and south Florida. Unpublished Ph.D. dissertation, Auburn University. Auburn, AL.

Butterfield, B.P., I. Fox, J. Garner, K. Carter, and J.B. Hauge. 2000. Geographic distribution: *Hemidactylus mabouia* (tropical gecko). Herpetol. Rev. 31:53.

Butterfield, B.P., B. Hauge, and W.E. Meshaka, Jr. 1993. The occurrence of *Hemidactylus mabouia* on the United States mainland. Herpetol. Rev. 24:111–112.

Butterfield, B.P., W.E. Meshaka, Jr., and C. Guyer. 1997. Nonindigenous amphibians and reptiles. *In* D. Simberloff, D.C. Schmitz, and T.C. Brown (eds.). Strangers in Paradise. Impact and Management of Nonindigenous Species in Florida. pp. 123–138. Island Press. Washington, DC.

Butterfield, B.P., W.E. Meshaka, Jr., and J.B. Hauge. 1994a. Two turtles new to the Florida Keys. Herpetol. Rev. 25:81.

Butterfield, B.P., W.E. Meshaka, Jr., and R.L. Kilhefner. 1994b. Two anoles new to Broward County, Florida. Herpetol. Rev. 25:77–78.

Butterfield, B.P. and J.B. Hauge. 2000. Geographic distribution: *Gekko gecko* (tokay gecko). Herpetol. Rev. 31:52.

Callahan, Jr., R.J. 1982. Geographic and ecological distribution of the lizard *Leiocephalus carinatus armouri* in Florida. Unpublished M.S. thesis, Univ. South Florida. Tampa, FL.

Campbell, T. 1999. Geographic distribution: *Osteopilus septentrionalis* (Cuban treefrog). Herpetol. Rev. 30:50–51.

Campbell, T.S. 1996. Northern range expansion of the brown anole (*Anolis sagrei*) in Florida and Georgia. Herpetol. Rev. 27:155–157.

Campbell, T.S. 2000. Analyses of the effects of an exotic lizard (*Anolis sagrei*) on a native lizard (*Anolis carolinensis*) in Florida, using islands as experimental units. Unpublished Ph.D. dissertation, Univ. Tennessee. Knoxville, TN.

Campbell, T.S. and G.P. Gerber. 1996. *Anolis sagrei* (brown anole). Saurophagy. Herpetol. Rev. 27:200.

Campbell, T.S. and J.T. Hammontree. 1995. Geographic distribution: *Anolis sagrei* (brown anole). Herpetol. Rev. 26:107.

Carr, A.F., Jr. 1939. A geckonid lizard new to the fauna of the United States. Copeia 1939:232.

Carr, A.F., Jr. 1940. A contribution to the herpetology of Florida. Univ. Florida Publ., Biol. Sci. Ser. 3:1–118

Christman, S.P., C.A. Young, S. Gonzalez, K. Hill, G. Navratil, and P. Delis. 2000. New records of amphibians and reptiles from Hardee County, Florida. Herpetol. Rev. 31:116–117.

Cochran, P.A. 1990. Geographic distribution: *Anolis sagrei* (brown anole). Herpetol. Rev. 21:22.

Collette, B.B. 1961. Correlations between ecology and morphology in anoline lizards from Havana, Cuba and southern Florida. Bull. Mus. Comp. Zool. 125:137–162.

Collins, J.T. and K.J. Irwin. 2001. Geographic distribution: *Hemidactylus turcicus* (Mediterranean gecko). Herpetol. Rev. 32:276.

Collins, J.T. and T.W. Taggart. 2002. Standard Common and Current Scientific Names for North American Amphibians, Turtles, Reptiles, and Crocodilians. Fifth Edition. Publication of The Center for North American Herpetology. Lawrence, KS.

Conant, R. 1975. A Field Guide to Reptiles and Amphibians of Eastern and Central North America. Houghton Mifflin. Boston, MA.

Conant, R. and J.T. Collins. 1991. A Field Guide to Reptiles and Amphibians of Eastern and Central North America. 3rd ed. Houghton Mifflin. Boston, MA.

Conant, R. and J.T. Collins. 1998. A Field Guide to Reptiles and Amphibians of Eastern and Central North America. 3rd ed., expanded. Houghton Mifflin. Boston, MA.

Cope, E.D. 1875. Check-list of North American Batrachia and Reptilia. Bull. U.S. Nat. Mus. 1:1–104.

Cope, E.D. 1889. The Batrachia of North America. Bull. U.S. Nat. Mus. No. 34.

Corwin, C.M., A.V. Linzey, and D.W. Linzey. 1977. Geographic distribution: *Anolis sagrei sagrei* (Cuban brown anole). Herpetol. Rev. 8:84.

Cox, J., R. Kautz, M. MacLaughlin, and T. Gilbert. 1994. Closing the Gaps in Florida's Wildlife Habitat Conservation System. Florida Game and Freshwater Fish Commission, Office of Environmental Services. Tallahassee, FL.

Crawford, D.M. and L.A. Somma. 1993a. Geographic distribution: *Ramphoty-phlops braminus* (Brahminy blind snake). Herpetol. Rev. 24:68.

Crawford, D.M. and L.A. Somma. 1993b. Geographic distribution: *Hemidactylus garnotii* (Indo-pacific gecko). Herpetol. Rev. 24:108–109.

Crews, D., J.E. Gustafson, and R.R. Tokarz. 1983. Psychobiology of Parthnogenesis. *In* R.B. Huey, E.R. Pianka and T.W. Schoener (eds.). Lizard Ecology. Studies of a Model Organism, pp 205–231. Harvard University Press. Cambridge.

Criscione, C.D., N.J. Anderson, T. Campbell, and B. Quinn. 1998. Geographic distribution: *Hemidactylus mabouia* (tropical gecko). Herpetol. Rev. 29:248.

Crump, M.L. 1986. Cannibalism by younger tadpoles: another hazard of metamorphosis. Copeia 1986:1007–1009.

Dalrymple, G.H. 1980. Comments on the density and diet of a giant anole, *Anolis equestris*. J. Herpetol. 14:412–415.

Dalrymple, G.H. 1988. The herpetofauna of Long Pine Key, Everglades National Park, in relation to vegetation and hydrology. *In* R.C. Szaro, K.E. Severson and D.R. Patton, (eds.). Proc. Symp. Management of Amphibians, Reptiles, and Small Mammals in North America, pp. 72–86. U.S. Forest Service Gen. Tech. Rep. RM-166. Fort Collins, CO.

Davis, W.K. 1974. The Mediterranean gecko, *Hemidactylus turcicus,* in Texas. J. Herpetol. 8:77–80.

Deckert, R.F. 1921. Amphibian notes from Dade County, Florida. Copeia 1921:20–23.

Delorey, C.J. and H.R. Mushinsky. 1987. Geographic distribution: *Ramphoty-phlops bramina* (Brahminy blind snake). Herpetol. Rev. 18:56.

De Sola, C.R. 1934. *Phrynosoma* from Florida. Copeia 1934:190.

Doan, T.M. 1996. Basking behavior of two *Anolis* lizards in south Florida. Florida Scientist, 59:16–19.

Drake, J.A., H.A. Mooney, F. Di Castri, R.H. Groves, F.J. Kruger, M. Rejmanek, and M. Williamson (eds.). 1989. Biological Invasions: A Global Perspective. John Wiley and Sons Ltd. New York.

Duellman, W.E. and R.I. Crombie. 1970. *Hyla septentrionalis* Duméril and Bibron. Cuban treefrog. Cat. Am. Amphib. Reptiles, 92.1–92.4.

Duellman, W.E. and A. Schwartz. 1958. Amphibians and reptiles of southern Florida. Bull. Florida State Mus. 3:181–324.

Dunson, W.A. 1982. Low water vapor conductance of hard-shelled eggs of the gecko lizards *Hemidactylus* and *Lepidodactylus*. J. Exper. Zool. 219:337–379.

Dunson, W.A. and C.R. Bramham.1981. Evaporative water loss and oxygen consumption of three small lizards from the Florida keys: *Sphaerodactylus cinereus, S. notatus,* and *Anolis sagrei*. Physiol. Zool. 54:253–259.

Duquesnel, J. 1998. Keys invasion by alien lizards continues. Florida Department of Environmental Protection, Resource Management Notes 10:9.

Eason, G.W., Jr. and D.R. McMillan. 2000. Geographic distribution: *Hemidacty-lus turcicus* (Mediterranean gecko). Herpetol. Rev. 31:53.

Eggert, J. 1978. The invasion of the wish willy. Florida Wildlife 31:9–10.

Ehrig, R.W. 1990. Geographic distribution: *Ramphotyphlops braminus* (Brahminy blind snake). Herpetol. Rev. 21:41.

Ehrlich, P.R. 1989. Attributes of invaders and the invading processes: vertebrates. *In* J.A. Drake, H.A. Mooney, F. Di Castri, R.H. Groves, F.J. Kruger, M. Rejmanek, and M. Williamson, (eds.), Biological Invasions: A Global Perspective, pp. 315–328. John Wiley and Sons Ltd. New York.

Ellis, T.M. 1980. *Caiman crocodilus:* An established exotic in south Florida. Copeia 1980:152-154.

Enge, K.M. 1998. Herpetofaunal survey of an upland hardwood forest in Gadsden County, Florida. Florida Scientist 61:141–159.

Enge, K.M. and K.N. Wood. 1999–2000. A herpetofaunal survey of Chassahowitzka Wildlife Management Area, Hernando County, Florida. Herpetol. Nat. Hist. 7:117–144.

Epler, J.H. 1986. Geographic distribution: *Sphaerodactylus elegans* (Ashy gecko). Herpetol. Rev. 17:27.

Ernst, C.H. and C.W. Brown. 2000. Geographic distribution: *Ramphotyphlops braminus* (Brahminy blind snake) Herpetol. Rev. 31:256.

Florida Fish and Wildlife Conservation Commission. 1999–2002. Critter Questions, Florida's Exotic Wildlife: http://wld.fwc.state.fl.us/critters/exotics/exotics. asp.

Fowler, H.W. 1915. Cold-blooded vertebrates from Florida, the West Indies, Costa Rica, and eastern Brazil. Proc. Acad. Nat. Sci. Philadelphia 67:244–269.

Frank, N. and E. Ramus. 1995. A Complete Guide to Scientific and Common Names of Reptiles and Amphibians of the World. N G Publishing Inc. Pottsville, PA.

Frankenberg, E. 1982a. Social behavior of the parthenogenetic Indo-Pacific gecko, *Hemidactylus garnotii*. Z. Tierpsychol. 59:19–28.

Frankenberg, E. 1982b. Vocal behavior of the Mediterranean house gecko, *Hemidactylus* turcicus. Copeia 1982:770–775.

Frankenberg, E. 1984. Interactions between two species of colonizing house geckos, *Hemidactylus turcicus* and *Hemidactylus garnotii*. J. Herp. 18:1–7.

Funk, R.S. and D. Moll. 1979. Geographic distribution: *Anolis sagrei sagrei* (Cuban brown anole). Herpetol. Rev. 10:102.

Garman, S. 1887. On West Indian Iguanidae and on West Indian Scincidae in the collection of the Museum of Comparative Zoology at Cambridge, Mass., USA. Bull. Essex Inst. 19:25–50.

Gerber, G.P. and A.C. Echternacht. 2000. Evidence for asymmetrical intraguild predation between native and introduced *Anolis* lizards. Oecologia, 124:599–607.

Gibbons, J.W. (ed.). 1990. Life History and Ecology of the Slider Turtle. Smithsonian Institution Press. Washington, DC.

Godley, J.S., F.E. Lohrer, J.N. Layne, and J. Rossi. 1981. Distributional status of an introduced lizard in Florida: *Anolis sagrei*. Herpetol. Rev. 12:84–86.

Goff, C.C. 1935. An additional note on *Phrynosoma cornutum* in Florida. Copeia 1935: 45.

Goin, C.J. 1944. *Eleutherodactylus ricordii* at Jacksonville, Florida. Copeia 1944: 192.

Goin, C.J. 1947. Studies on the life history of *Eleutherodacytlus ricordii planirostris* (Cope) in Florida, with special reference to the local distribution of an allelomorphic color pattern. University of Florida Press. Gainesville, FL.

Günther, R., A.M. Bauer, and D. King. 1993. Geographic distribution: *Hemidactylus mabouia* (tropical house gecko). Herpetol. Rev. 24:66.

Harper, F. 1935. Records of the amphibians of the southeastern states. Amer. Midl. Natur. 16:275–310.

Harris, V.A. 1964. The Life of the Rainbow Lizard. Hutchinson Tropical Monographs. London, England.

Hauge, J.B. and B.P. Butterfield. 2000a. Geographic distribution: *Cosymbotus platyurus* (Asian house gecko). Herpetol. Rev. 31:52.

Hauge, J.B. and B.P. Butterfield. 2000b. Geographic distribution: *Leiocephalus carinatus armouri* (northern curlytail lizard). Herpetol. Rev. 31:53.

Hutchison, A.M. 1992. A reproducing population of *Trachemys scripta elegans* in southern Pinellas County, Florida. Herpetological Review 23:74–75.

Inger, R.F. and B. Greenberg, 1966. Ecological and competitive relations among three species of frog (genus *Rana*). Ecology 47:746–759.

Irwin, K.J. 1999. Geographic distribution: *Eleutherodactylus planirostris* (greenhouse frog). Herpetol. Rev. 30:106.

Jensen, J.B. 1994. Geographic distribution: *Phrynosoma cornutum* (Texas horned lizard). Herpetol. Rev. 25:165.

Jensen, J.B. 1995. Geographic distribution: *Hemidactylus turcicus* (Mediterranean gecko). Herpetol. Rev. 26:45.

Jensen, J.B. and J.G. Palis. 1995. Geographic distribution: *Eleutherodactylus planirostris* (greenhouse frog). Herpetol. Rev. 26:104.

King, F.W. 1958. Observations on the ecology of a new population of the Mediterranean gecko, *Hemidactylus turcicus,* in Florida. Quarterly J. Florida Acad. Sci. 21:317–318.

King, F.W. 1960. New populations of West Indian reptiles and amphibians in southeastern Florida. Quarterly J. Florida Acad. Sci. 23:71–73.

King, F.W. 1966. Competition between two south Florida lizards of the genus *Anolis*. Unpublished Ph.D. dissertation, Univ. Miami. Miami, FL.

King, F.W. and T. Krakauer, 1966. The exotic herpetofauna of southeastern Florida. Quarterly J. Florida Acad. Sci. 29:144–154.

Kluge, A.G. and M.J. Eckhardt. 1969. *Hemidactylus garnotii* Duméril and Bibron, a triploid all-female species of geckonid lizard. Copeia 1969:651–664.

Krakauer, T. 1968. The ecology of the neotropical toad, *Bufo marinus,* in south Florida. Herpetologica 24:214–221.

Krakauer, T. 1970. The invasion of the toads. Florida Nat. 43:12–14.

Kraus, F., E. W. Campbell, III, A. Allison, and T. Pratt. 1999. *Eleutherodactylus* frog introductions to Hawaii. Herpetol. Rev. 30:21–25.

Krysko, K.L. and F.W. King. 1999. Geographic distribution: *Osteopilus septentrionalis* (Cuban treefrog). Herpetol. Rev. 30:230–231.

Krysko, K.L. and F.W. King. 2000. Geographic distribution: *Eleutherodactylus planirostris* (greenhouse frog). Herpetol. Rev. 31:109.

Krysko, K.L. and A.T. Reppas. 1999. Geographic distribution: *Eleutherodactylus planirostris* (greenhouse frog). Herpetol. Rev. 30:106.

Krysko, K.L., J.N. Decker, and A.T. Reppas. 2000. Geographic distribution: *Ramphotyphlops braminus* (Brahminy blind snake). Herpetol. Rev. 31:256.

Lawson, R., P.G. Frank, and D.L. Martin. 1991. A gecko new to the United States herpetofauna, with notes on geckos of the Florida Keys. Herpetol. Rev. 22:11–12.

Layne, J.N. 1987. Geographic distribution: *Leiocephalus carinatus* (curly-tailed lizard). Herpetol. Rev. 18:20.

Layne, J.N., J.A. Stallcup, G.E. Woolfenden, M.N. McCauley, and D.J. Worley. 1977. Fish and Wildlife Inventory of the Seven-County Region Included in the Central Florida Phosphate Industry Areawide Environmental Impact Study). U.S. Dept. Commerce, Nat. Tech. Info. PB-278 456 Vol. 1.

Lazell, J.D., Jr. 1989. Wildlife of the Florida Keys: A Natural History. Island Press. Washington, DC.

Lee, D.S. 1969. Floridian herpetofauna associated with cabbage palms. Herpetologica 25:70–71.

Lee, J.C. 1985. *Anolis sagrei* in Florida: phenetics of a colonizing species I. meristic characters. Copeia 1985:182–194.

Lee, J.C. 1987. *Anolis sagrei* in Florida: phenetics of a colonizing species II. morphometric characters. Copeia 1987:458–469.

Lee, J.C. 1992. *Anolis sagrei* in Florida: phenetics of a colonizing species III. West Indian and Middle American comparisons. Copeia 1992:942–954.

Lee, J.C. 1996. The Amphibians and Reptiles of the Yucatan Peninsula. Cornell Univ. Press. Ithaca, NY.

Lee, J.C., D. Clayton, S. Eisenstein, and I. Perez. 1989. The reproductive cycle of *Anolis sagrei* in southern Florida. Copeia 1989:930–937.

Lieb, C.S. D.G. Buth, and G.C. Gorman. 1983. Genetic differentiation in *Anolis sagrei:* a comparison of Cuban and introduced Florida populations. J. Herp. 17:90–94

Lindsay, C.R. and J.H. Townsend. 2001. Geographic distribution: *Hemidactylus garnotii* (Indo-pacific gecko). Herpetol. Rev. 32:193.

Lips, K.R. and J.N. Layne. 1989. Vertebrates associated with gopher tortoise (*Gopherus polyphemus*) burrows in four habitats in south-central Florida. Florida Scientist 52:20 (abstr.).

Loftus, W.F. and R. Herndon. 1984. Reestablishment of the coqui, *Eleutherodactylus coqui* Thomas, in southern Florida. Herpetol. Rev. 15:23.

Love, B. 2000. *Gekko gecko* (tokay gecko). Predation. Herpetol. Rev.31:174.

Love, W.B. 1978. Observations on the herpetofauna of Key West, Florida, with special emphasis on the rosy rat snake. Bull. Georgia Herpetol. Soc. 4: 3–8.

Love, W.B. 1995. *Osteopilus septentrionalis* (Cuban treefrog). Predation. Herpetol. Rev. 26:201–202.

McCoy, C.J. 1971. Geographic distribution: *Hemidactuylus turcicus* (Mediterranean gecko). Herpetol. Rev. 3:89.

McCoy, C.J. 1972. Geographic distribution: *Hemidactylus garnotii* (Indo-pacific gecko). Herpetol. Rev. 4:23.

Means, D.B. 1990a. Geographic distribution: *Anolis sagrei* (brown anole). Herpetol. Rev. 21:96.

Means, D.B. 1990b. Geographic distribution: *Hemidactylus turcicus* (Mediterranean gecko). Herpetol. Rev. 21:96.

Means, D.B. 1996a. Geographic distribution: *Anolis sagrei* (brown anole). Herpetol. Rev. 27:151–152.

Means, D.B. 1996b. Geographic distribution: *Gekko gecko* (tokay gecko). Herpetol. Rev. 27:152.

Means, D.B. 1996c. Geographic distribution: *Hemidactylus turcicus* (Mediterranean gecko). Herpetol. Rev. 27:152.

Means, R.C. 1999. Geographic distribution: *Hemidactylus turcicus* (Mediterranean gecko). Herpetol. Rev. 30:52.

Meshaka, W.E., Jr. 1993. Hurricane Andrew and the colonization of five invading species in south Florida. Florida Scientist 56:193–201.

Meshaka, W.E., Jr. 1994a. Giant toad eaten by a red-shouldered hawk. Florida Field Natur. 22:54–55.

Meshaka, W.E., Jr. 1994b. Reproductive cycle of the Indo-Pacific gecko (*Hemidactylus garnotii*) in south Florida. Florida Scientist 57:6–9.

Meshaka, W.E., Jr. 1994c. Geographic distribution: *Ramphotyphlops braminus* (Brahminy blind snake). Herpetol. Rev. 25:34.

Meshaka, W.E., Jr. 1995. Reproductive cycle and colonization ability of the Mediterranean gecko (*Hemidactylus turcicus*) in south-central Florida. Florida Scientist 58:10–15.

Meshaka, W.E., Jr. 1996a. Vagility and the Florida distribution of the Cuban treefrog (*Osteopilus septentrionalis*). Herpetol. Rev. 27:37–40.

Meshaka, W.E., Jr. 1996b. *Osteopilus septentrionalis* (Cuban treefrog): maximum size. Herpetol. Rev. 27:74.

Meshaka, W.E., Jr. 1996c. Retreat and habitat use of the Cuban treefrog, *Osteopilus* septentrionalis: implications for successful colonization in Florida. J. Herpetol. 30:443-445.

Meshaka, W.E., Jr. 1996d. Occurrence of the nematode *Skrjabinoptera scelopori* in the Cuban treefrog (*Osteopilus septentrionalis*): mainland and island comparisons. *In* Powell, R. and R.W. Henderson (eds.), Contributions to West Indian Herpetology: A Tribute to Albert Schwartz, pp 271–276. Contributions to Herpetology, volume 12. Society for the Study of Amphibians and Reptiles.

Meshaka, W.E., Jr. 1996e. Anuran Davian behavior: a Darwinian dilemma. Florida Scientist 59:74–75.

Meshaka, W.E., Jr. 1996f. Theft or cooperative foraging in the barred owl? Florida Field Natur. 24:15.

Meshaka, W.E., Jr. 1999a. The herpetofauna of the Doc Thomas House in South Miami, Florida. Florida Field Natur. 27:121–123.

Meshaka, W.E., Jr. 1999b. The herpetofauna of the Kampong. Florida Scientist. 62:153–157.

Meshaka, W.E., Jr. 1999c. Research and thoughts on the knight anole (*Anolis equestris*) in southern Florida. The *Anolis* Newsletter V:86–88.

Meshaka, W.E., Jr. 2000. Colonization dynamics in two exotic geckos (*Hemidactylus garnotii* and *H. mabouia*) in Everglades National Park. J. Herpetol. 34: 163–168.

Meshaka, W.E., Jr. 2001. The Cuban Treefrog in Florida: Life History of a Successful Colonizing Species. University Press of Florida. Gainesville, FL.

Meshaka, W.E., Jr. and B. Ferster. 1995. Two species of snakes prey on Cuban treefrogs in southern Florida. Florida Field Natur. 23:97–98.

Meshaka, W.E., Jr. and K.P. Jansen. 1997. *Osteopilus septentrionalis* (Cuban treefrog): predation. Herpetol. Rev. 28:147–148.

Meshaka, W.E., Jr. and J. Lewis. 1994. *Cosymbotus platyurus* in Florida: ten years of stasis. Herpetol. Rev. 25:127.

Meshaka, W.E., Jr. and B.A. Moody. 1996. The Old World tropical house gecko (*Hemidactylus mabouia*) on the Dry Tortugas. Florida Scientist 59:115–117.

Meshaka, W.E., Jr. and K.G. Rice. In Review. The knight anole (*Anolis equestris*): Ecology of a successful colonizing species in extreme southern mainland Florida *In* W.E. Meshaka, Jr. and K.J. Babbitt (eds.), Status and Conservation of Florida Amphibians and Reptiles. Krieger. Malabar, FL.

Meshaka, W.E., Jr., B.P. Butterfield, and B. Hauge. 1994a. *Hemidactylus mabouia* as an established member of the Florida herpetofauna. Herpetol. Rev. 25:80–81.

Meshaka, W.E., Jr., B.P. Butterfield, and B. Hauge. 1994b. *Hemidactylus frenatus* established in Florida. Herpetol. Rev. 25:127–128.

Meshaka, W.E., Jr., B.P. Butterfield, and J.B. Hauge. 1994c. Geographic distribution. *Hemidactylus mabouia* (tropical house gecko). Herpetol. Rev. 25:165.

Meshaka, W.E. Jr., B.P. Butterfield, and J.B. Hauge. 1994d. Reproductive notes on the introduced gecko *Hemidactylus mabouia* in southern Florida. Herpetol. Nat. Hist. 2:109–110.

Meshaka, W.E., Jr., R.M. Clouse, B.P. Butterfield, and J.B. Hauge. 1997a. The Cuban green anole, *Anolis porcatus:* a new anole established in Florida. Herpetol. Rev. 28:101–102.

Meshaka, W.E., Jr., R. Clouse, and L. MacMahon. 1997b. Diet of the tokay gecko (*Gekko gecko*) in southern Florida. Florida Field Natur. 25:105–107.

Meshaka, W.E., Jr., W.B. Loftus, and T.M. Steiner. 2000. The herpetofauna of Everglades National Park. Florida Scientist 63:84–103.

Meshaka, W.E., Jr., R. Snow, O.L. Bass, Jr., and W.B. Robertson, Jr. In Press. Occurrence of wildlife on tree islands in the southern Everglades *In* A. Van Der Valk and F. Sklar (eds.), Tree Islands of the Everglades, pp—- —. Kluwer Press.

Meylan, P.A. 1977a. Geographic distribution: *Anolis sagrei* (brown anole). Herpetol. Rev. 8:39.

Meylan, P.A. 1977b. Geographic distribution: *Hemidactylus turcicus* (Mediterranean gecko). Herpetol. Rev. 8:39.

Mitchell, J.C. and W.B. Hadley. 1980. Geographic distribution: *Hemidactylus garnotii* (Indo-pacific gecko). Herpetol. Rev. 11:80.

Miyamoto, M.M., M.P. Hayes, and M.R. Tennant. 1986. Biochemical and morphological variation in Floridian populations of the bark anole (*Anolis distichus*). Copeia 1986:76-86.

Myers, S. 1977. Geographic distribution: *Osteopilus septentrionalis* (Cuban treefrog). Herpetol. Rev. 8:38.

Myers, S. 1978a. Geographic distribution: *Hemidactylus turcicus* (Mediterranean gecko). Herpetol. Rev. 9:62.

Myers, S. 1978b. Geographic distribution: *Hemidactylus turcicus* (Mediterranean gecko). Herpetol. Rev. 9:107.

Myers, S. 1978c. Geographic distribution: *Anolis sagrei* (brown anole). Herpetol. Rev. 9:107-108.

Myers, S. 1979. Geographic distribution: *Hemidactylus garnoti* (Indo-pacific gecko). Herpetol. Rev. 10:102–103.

Neill, W.T. 1951. A bromeliad herpetofauna in Florida. Ecology 31:140–143.

Neill, W.T. 1957. Historical biogeography of present-day Florida. Bull. Florida State Mus. 2:175–220.

Nelson, D.H. and J.D. Carey. 1993. Range extension of the Mediterranean gecko (*Hemidactylus turcicus*) along the northeastern gulf coast of the United States. Northeast Gulf Science, 13:53–58.

Nicholson, K.E., A.V. Paterson, and P.M. Richards. 2000. *Anolis sagrei* (brown anole). cannibalism. Herpetol. Rev. 31:173–174.

Noonan, B. 1995. Geographic distribution: *Anolis equestris* (knight anole). Herpetol. Rev. 26:209.

Ober, L.D. 1973. Introduction of the Haitian anole, *Anolis cybotes*, in the Miami area. HISS News-Journal, 1:99.

Oliver, J.A. 1949. The peripatetic toad. Nat. Hist. 58:23–33.

Oliver, J.A. 1950a. Natural History of North American Amphibians and Reptiles. D. Van Nostrand Co., Inc. Princeton, NJ.

Oliver, J.A. 1950b. *Anolis sagrei* in Florida. Copeia 1950:55–56.

Oliver, J.H., Jr., M.P. Hayes, J.E. Keirans, and D.R. Lavender. 1993. Establishment of the foreign parthenogenetic tick *Amblyomma rotundatum* (Acari: Ixodidae) in Florida. J. Parasitol. 79:786–790.

Owen, R.D., D.T. Bowman, Jr., and S.A. Johnson. 1998. Geographic distribution. *Ramphotyphlops braminus* (Brahminy blind snake). Herpetol. Rev. 29:115.

Pianka, E. R. 1986. Ecology and Natural History of Desert Lizards. Analyses of the Ecological Niche and Community Structure. Princeton University Press. Princeton, NJ.

Pough, F.H., R.M. Andrews, J.E. Cadle, M.L. Crump, A.H. Savitzky, and K.D. Wells. 2001. Herpetology. 2nd Edition Prentice Hall, Upper Saddle River, NJ.

Ranzzetti, E.R. and C.A. Msuya. 2002. Field Guide to the Amphibians and Reptiles of Arusha National Park (Tanzania). Tanzania National Parks (TANAPA).

Reichard, S.M. and H.M. Stevenson. 1964. Records of *Eleutherodactylus ricordi* at Tallahassee. Florida Naturalist 37:97.

Reppas, A.T. 1999. Geographic distribution: *Hemidactylus garnotii* (Indo-pacific gecko). Herpetol Rev. 30:110.

Reppas, A.T., K.L. Krysko, C.L. Sonberg, and R.H. Robins. 1999. Geographic distribution: *Anolis distichus* (bark anole). Herpetol. Rev. 30:51.

Riemer, W.J. 1958. Giant toads of Florida. Quarterly J. Florida Acad. Sci. 21: 207–211.

Rossi, J.V. 1981. *Bufo marinus* in Florida: some natural history and its impact on native vertebrates. Unpublished M.A. thesis, Univ. South Florida. Tampa, FL.

Rossi, J.V. 1983. The use of olfactory cues by *Bufo marinus*. J. Herpetol. 17: 72–73.

Ruibal, R. 1964. An annotated checklist and key to the anoline lizards of Cuba. Bull. Mus. Comp. Zool. 130:473–520.

Salzburg, M.A. 1984. *Anolis sagrei* and *Anolis cristatellus* in southern Florida: a case study in interspecific competition. Ecology, 65:14–19.

Savage, J.M. 1954. Notulae herpetologicae 1–7. Trans. Kansas Acad. Sci. 57: 326–334.

Schoener, T. and A. Schoener. 1980. Densities, sex ratios, and population structure in four species of Bahamian *Anolis* lizards. J. Animal Ecol. 49:29–53.

Schwartz, A. 1952. *Hyla septentrionalis* Duméril and Bibron on the Florida mainland. Copeia 1952:117–118.

Schwartz, A. 1971. *Anolis distichus*. Cat. Amer. Amphib. Reptiles, 108.1–108.4.

Schwartz, A. 1974. *Eleutherodactylus planirostris* (Cope). Cat. Amer. Amphib. Reptiles 154.1–154.4.

Schwartz, A. and R.W. Henderson. 1991. Amphibians and Reptiles of the West Indies: Descriptions, Distributions, and Natural History. University of Florida Press. Gainesville, FL.

Schwartz, A. and R. Thomas. 1975. A check-list of West Indian amphibians and reptiles. Carnegie Mus. Nat. Hist. Spec. Pub. 1:1–216.

Seigel, B.J., N.A. Seigel, and R.A. Seigel. 1999. Geographic distribution: *Anolis cristatellus* (Puerto Rican crested anole). Herpetol. Rev. 30:173.

Simberloff, D., D.C. Schmitz, and T.C. Brown (eds.). 1997. Strangers in Paradise. Island Press. Washington, DC.

Skermer, G.M. 1939. Notes on *Eleutherodactylus ricordii*. Copeia 1939:107–108.

Smith, H.M. and A.J. Koehler. 1978. A survey of herpetological introductions in the United States and Canada. Trans. Kansas Acad. Sci. 80:1–24.

Smith, H.M. and R.H. McCauley. 1948. Another new anole from south Florida. Proc. Biol. Soc. Washington 61:159–166.

Smith, R.E. 1983. Geographic distribution: *Hemidactylus garnoti* (Indo-pacific gecko). Herpetol. Rev. 14:84.

Somma, L.A. and D.M. Crawford. 1993. Geographic distribution: *Osteopilus septentrionalis* (Cuban treefrog). Herpetol. Rev. 24:153.

Steiner, T.M. and L.T. McLamb. 1982. Geographic distribution: *Hemidactylus garnoti* (Indo-pacific gecko). Herpetol. Rev. 13:25.

Stejneger, L. 1922. Two geckos new to the fauna of the United States. Copeia 1922:56.

Stevenson, D. and D. Crowe. 1992a. Geographic distribution: *Bufo marinus* (giant toad). Herpetol. Rev. 23:85.

Stevenson, D. and D. Crowe. 1992b. Geographic distribution: *Anolis sagrei* (brown anole). Herpetol. Rev. 23:89.

Stevenson, D. and D. Crowe, 1992c. Geographic distribution: *Hemidactylus garnoti* (Indo-pacific gecko). Herpetol. Rev. 23:90.

Stevenson, H.M. 1976. Vertebrates of Florida. University of Florida Press, Gainesville, FL.

Timmerman, W.W., J.B. Miller, and C.V. Tamborski. 1994. The herpetofauna of Jonathan Dickinson State Park, Martin County, Florida. Florida Department of Environmental Protection, Division of Recreation and Parks, Hobe Sound, Florida, USA. Final Report Project No. 7618.

Townsend, J.H. and C.R. Lindsay. 2001. Geographic distribution: *Anolis sagrei* (brown anole). Herpetol. Rev. 32:193.

Townsend, J.H. and A.Y. Reppas. 2001. Geographic distribution: *Hemidactylus turcicus* (Mediterranean gecko). Herpetol. Rev. 32:193.

Van Hyning, O.C. 1933. Batrachia and Reptilia of Alachua County, Florida. Copeia 1933:3–7.

Voss, R. 1975. Notes on the introduced gecko *Hemidactylus garnoti* in south Florida. Florida Scientist 38:174.

Wallach, V., G.S. Jones, and R.R. Kunkel. 1991. Geographic distribution: *Ramphotyphlops braminus* (Braminy blind snake). Herpetol. Rev. 22:68.

Watkins-Colwell, G.J. and K.A. Watkins-Colwell. 1995a. Geographic distribution: *Anolis distichus* (bark anole). Herpetol. Rev. 26:44.

Watkins-Colwell, G.J. and K.A. Watkins-Colwell. 1995b. Geographic distribution: *Hemidactylus mabouia* (tropical house gecko). Herpetol. Rev. 26:45.

Watkins-Colwell, G.J. and K.A. Watkins-Colwell. 1995c. Geographic distribution: *Ramphotyphlops braminus* (Brahminy blind snake). Herpetol. Rev. 26:210.

Weigl, G.L., R.G. Domey, and W.R. Courtenay, Jr. 1969. Survival and range expansion of the curly-tailed lizard, *Leiocephalus carinatus armouri,* in Florida. Copeia 1969: 841–842.

Wilson L.D. and L. Porras. 1983. The ecological impact of man on the South Florida herpetofauna. Univ. Kansas Mus. Nat. Hist., Spec. Pub. No. 9. Lawrence, KS.

Wise, M.A. 1993. Geographic distribution: *Hemidactylus turcicus* (Mediterranean gecko). Herpetol. Rev. 24:109.

Wray, K. and R. Owen. 1999. New records of amphibians and reptiles for Nassau County, Florida. Herpetol. Rev. 30:237–238.

Wygoda, M.L. and J.R. Bain. 1980. Geographic distribution: *Anolis sagrei* (Cuban brown anole). Herpetol. Rev. 11:115.

Zhao, E. and K. Adler. 1993. Herpetology of China. Contributions to Herpetology No. 10, Society for the Study of Amphibians and Reptiles.